Studies in Universal Logic

This series is devoted to the universal approach to logic and the development of a general theory of logics. It covers topics such as global set-ups for fundamental theorems of logic and frameworks for the study of logics, in particular logical matrices, Kripke structures, combination of logics, categorical logic, abstract proof theory, consequence operators, and algebraic logic. It includes books with historical and philosophical discussions about the nature and scope of logic. Three types of books will appear in the series: graduate textbooks, research monographs, and volumes with contributed papers.

Studies in Universal Logic

This series is devoted to the universal approach to logic and the development of a general theory of logics. It covers topics such as global set-ups for fundamental theorems of logic and frameworks for the study of logics, in particular logical matrices, Kripke structures, combination of logics, categorical logic, abstract proof theory, consequence operators, and algebraic logic. It includes also books with historical and philosophical discussions about the nature and scope of logic. Three types of books will appear in the series: graduate textbooks, research monographs, and volumes with contributed papers.

Olivier Gasquet • Andreas Herzig • Bilal Said •
François Schwarzentruber

Kripke's Worlds

An Introduction to Modal Logics via Tableaux

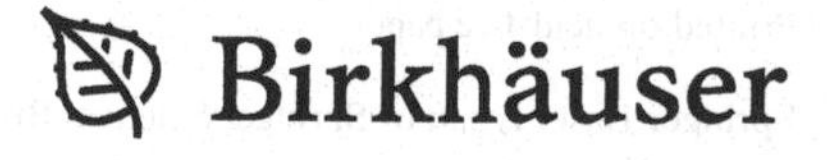

Olivier Gasquet
Institut de Recherche en Informatique de Toulouse (IRIT)
Université Paul Sabatier
Toulouse, France

Andreas Herzig
Institut de Recherche en Informatique de Toulouse (IRIT)
Université Paul Sabatier
Toulouse, France

Bilal Said
Institut de Recherche en Informatique de Toulouse (IRIT)
Université Paul Sabatier
Toulouse, France

François Schwarzentruber
Institut de Recherche en Informatique de Toulouse (IRIT)
Université Paul Sabatier
Toulouse, France

ISBN 978-3-7643-8503-3
ISBN 978-3-7643-8504-0 (eBook)
DOI 10.1007/978-3-7643-8504-0
Springer Basel Heidelberg New York Dordrecht London

Library of Congress Control Number: 2013955232

Mathematics Subject Classification (2010): 03-XX, 03B42, 03B44, 03B45

Printed on acid-free paper

Springer Basel is part of Springer Science+Business Media (www.springer.com)

Preface

In classical logic—that is, in propositional logic and in first-order logic—, formulas are either true or false. Modal logic takes a closer look and allows one to distinguish formulas that are true but could be false from formulas that are not only true, but necessarily so. In the same vein, it allows one to distinguish formulas that are false but possibly true from formulas that are not only false, but necessarily so, i.e., impossible.

Aristotle in his *Organon* made exactly the above distinctions, and the concepts of necessity and possibility have interested philosophers since. However, there was no clear and simple semantics before the end of the 1950s. At that time, Saul Kripke popularised the idea that necessity of a formula should be interpreted as *truth of that formula in all possible worlds*; and likewise, possibility of a formula as *truth in some possible world* [Kri63]. Kripke's idea can be traced back to Leibniz and more recently Carnap. His groundbreaking contribution was to refine that interpretation: he proposed that a formula A has to be evaluated relative to a given possible world w and that A is necessarily true at w if and only if A is true in all possible worlds *that are accessible from* w. So there might be worlds that are possible but inaccessible. While similar proposals were independently made by Marcel Guillaume, Stig Kanger, Jaakko Hintikka, and others, possible worlds semantics became known under the name "Kripke semantics".

What are possible worlds? For a philosopher this is a subject of debate, with issues such as: do they really exist? what kind of possibility are we talking about? etc. The picture is clear in theoretical computer science: possible worlds are viewed as states in which a computer program might be, and the relation of accessibility is viewed as a *transition relation between states*. Quite differently, in semantic web ontologies possible worlds are objects from some domain, and the accessibility relations are *relations between objects*. Possible worlds are *nodes in a graph*: node w_2 is accessible from node w_1 via the accessibility relation if there is an edge (an arrow) from w_1 to w_2 that is labelled by that relation.

Beyond Kripke's original formal analysis of the concepts of necessity and possibility, possible worlds models quickly turned out to be a flexible tool to investigate a whole family of other concepts.

- Temporal concepts such as 'always' and 'sometimes,' 'henceforth' and 'eventually,' 'next,' 'until' and 'since,' 'before' and 'after' can be naturally interpreted with accessibility relations. These relations are typically orders, in particular linear orders. This was first proposed by Arthur Prior [Pri57, Pri67].
- What is necessarily or possibly true after the execution of a (possibly nondeterministic) computer program π can be described by associating transition relations between states to π. This was initiated by Vaughn Pratt, after earlier work by Andrzej Salwicki [Pra76].
- Knowledge and belief of an agent I can be interpreted as truth in all worlds that are possible for I. This was first proposed by Jaakko Hintikka [Hin62].
- Deontic concepts such as obligation and permission can be interpreted as truth in all (resp. some) ideal worlds. Building on earlier work by G.H. von Wright [vW51], such an analysis was advocated by Stig Kanger and Jaakko Hintikka.
- Motivational concepts such as goals and intentions of an individual I can be interpreted as truth in all worlds that are preferred by I. This was studied among others by Philip Cohen and Hector Levesque [CL90a, CL90b].
- Conditionals of the form "If A then B" were interpreted as truth of B in all those worlds where A is true and that are closest to the actual world. This was studied by Robert Stalnaker and David Lewis [Sta68, Lew73].

All these concepts are called *modalities*. That is the wide sense of the term; in the narrow sense, there are only two modalities: necessity and possibility.

When investigating the modal logic of a particular concept such as belief one typically starts by analysing the *logical form* of that concept. For example, the object of an agent's belief is a formula, and an agent's belief that a formula is true is itself a formula. One then defines a *logical language* containing one or more modal operators: a single operator of belief if one only considers one agent, or several operators of belief, one per agent. The formulas of that language can then be interpreted in terms of possible worlds and accessibility relations between possible worlds. Each particular concept one wishes to investigate has specific properties. For example, the temporal concept of a formula being eventually true is interpreted by an accessibility relation that is both reflexive and transitive: if A is true now then it can be said that A is eventually true; and if it is eventually true that A is eventually true, then A is eventually true; in contrast, the temporal 'next' ('at the next state') relation is neither reflexive nor transitive.

To design a logic amounts to identifying the list of constraints that the accessibility relations should satisfy. Formulas of a modal language are therefore interpreted in particular *classes of models*. For example the language with the modal operator 'eventually' is interpreted in the class of Kripke models whose accessibility relation for 'eventually' is reflexive and transitive.

In all these proposals, possible worlds are viewed as complete descriptions of a state of affairs. In a different tradition, possible worlds are viewed as objects of the world (just as in first-order logic), and the formula A is viewed as a property of the object under concern. Then 'necessarily A' expresses a restricted quantification: all objects that are related to the object currently under concern have property A. This view leads to the definition of knowledge representation languages that generalise relational databases and semantic networks. Their advantage over first-order logic is that—just as many other modal logics—they have good mathematical properties. They are in particular said to be "Robustly decidable" [Var96]: More precisely, it is not only decidable whether a given formula A is true in a given world of a given Kripke model, but also whether for a given formula A, a given class of Kripke models contains a model such that A is true in some world of that model. The former reasoning problem is called the model checking problem; the latter is called the satisfiability problem.

Why Did We Write This Book?

Why did we choose to write this book? Several textbooks on modal logics already exist, both at the introductory and at the more advanced level.

Almost all introductory texts start with syntax and semantics, and then focus on Hilbert-style axiomatizations. Examples of such books are the classical [HC68, HC84] and [Che80], and the more recent [CP09]. All of these devote relatively little attention to the above-mentioned reasoning tasks of model checking and satisfiability checking, nor are they geared towards automatic procedures. Moreover, the first three of them are almost exclusively about *monomodal* logics: modal logics having only one modal operator of necessity and one modal operator of possibility. They do not account for logics that are about more than one concept, such as logics of knowledge and action, logics of knowledge and obligation, or logics of belief and time.

There exist more advanced textbooks that cover not only semantics and Hilbert axiomatizations, but also currently known decidability and complexity results. Among these books there are [CZ97, BdRV01, GKWZ03, BBW06, GSS03]. However, all these books contain quite demanding material going far beyond introductory textbooks. The same can be said for textbooks about particular modal logics such as the logic of programs [HKT00], the logics of time [FGV05], or the logics of ontologies [BCM+03].

The present book aims at filling this gap: our aim is to introduce the reader to the most important modal logics with multiple modalities, and we would like to do so from the perspective of the associated automated reasoning tasks. To that end we concentrate on the most general and powerful reasoning method for modal logics: tableaux systems.

The central idea of the tableaux method is: try to build a Kripke model for a given formula by applying the truth conditions. More than 40 years ago, Melvin Fitting showed how to extend tableaux systems from classical logic to modal logics. He used these systems in his textbook order to introduce the basic modal logics in a systematic way [Fit83]. It has to be noted that Fitting exclusively treated monomodal logics; moreover, the tableaux systems were presented in the way they were supposed to be used at that time, namely in order to write down proofs with paper and pencil.

Since the 1990s there has been a trend to design and implement tableaux systems on computers. Prominent examples are the Logic Workbench LWB [HSZ96] and several tableaux provers for description logics [TH06]. For an up-to-date overview we refer the reader to a webpage that is maintained by Renate Schmidt at the University of Manchester.[1] Our motivation to write this book was to use implemented tableaux systems in order to provide a way to gently introduce the reader to the wealth of currently existing modal logics. However, the above-mentioned implementations are all dedicated to either just one modal logic, or to a small family thereof. In contrast, a general introduction to modal logics requires a general and versatile piece of software that is able to account for the numerous existing modal logics. That set being infinite, what is needed is a generic tool that can be easily instantiated in order to implement a given modal logic—possibly a new logic that had never been considered before. There exist only three generic tableaux provers: the Tableaux Workbench TWB[2] [AG09], the Tableau Prover Generator MetTeL2[3] [TSK12] and the vehicle of our book: LoTREC.

The tableaux method is usually presented in a way that comes very close to that of Gentzen sequent systems. The latter can be used to build proofs that take the form of trees, breaking down a given input formula connective by connective. Such sequent systems can be defined quite straightforwardly for many basic modal logics; by and large, those whose models can be identified with trees. However, Kripke models are not limited to trees, and both sequent and tableaux systems for logics with e.g. symmetric accessibility relations are more difficult to design: somewhat in opposition with Gentzen's spirit they typically require side conditions, leading to a cumbersome meta-linguistic machinery. Contrasting with almost all existing tree-based tableaux systems, we are going to work on graphs in order to get around that difficulty. Our presentation of tableaux is therefore much closer to Kripke models.

Staying close to Kripke models has a second advantage: it allows one to introduce at the same time the semantics of modal logics on the one hand, and tableaux proof systems on the other.

There is another difference with traditional presentations of tableaux systems: just as in the case of sequent systems, most of the explanations are in terms of the search for a proof of validity of a formula that is done by refuting its negation. In contrast, we focus on the construction of a model for the input formula: for us, a tableaux system is a *model construction* procedure.

The price to pay for generality and simplicity of use is lack of efficiency: while LoTREC performs reasonably well on the basic modal logics, it is outperformed by more specialized tableaux implementations... when run on the logics the latter are designed for. In contrast, when one wants to implement a new tableaux system then LoTREC provides a simple and generic language to do so, while existing implementations are difficult to adapt and require diving into the programming language code in which the tableau prover is implemented, such as C++ in the case of the LWB.

[1] http://www.cs.man.ac.uk/~schmidt/tools.

[2] http://twb.rsise.anu.edu.au.

[3] http://www.mettel-prover.org.

To sum it up, the aim of this book is to provide a gentle introduction to multimodal logics via tableaux systems. It should enable readers that do not have any knowledge beyond propositional logic to learn what modal logics are about, which properties they have and how they can be put to work; in particular, we hope that it will allow them to implement and play with existing or new modal logics. Most importantly, they should be able to do so *without any knowledge or skills in programming*. Our tool in this enterprise is LoTREC. In each of the chapters in the book the reader may turn into a LoTREC user in order to actively explore the functioning and the programming of tableaux systems.

LoTREC: An Ubiquitous Tool in This Book

LoTREC is a free software that is distributed by the *Institut de recherche en informatique de Toulouse* (IRIT) and that is accessible at the following webpage:

http://www.irit.fr/Lotrec

While the name LoTREC refers to the French painter Henri de Toulouse Lautrec, it can also be understood as the acronym of the—admittedly clumsy—"**Lo**gical **T**ableaux **R**esearch **E**ngine **C**ompanion."

The language of LoTREC has three basic concepts: logical connectives, tableau rules and strategies.

- The user of LoTREC can define his own logical language by defining logical connectives of any arity: one, two, or more.
- A *tableau rule* is of the form "If *condition* then *action*" and enables construction of Kripke models.
- A *strategy* describes in which order the rules are applied.

LoTREC is *generic*: by defining unique connectives, rules and strategies the LoTREC user can implement a tableau proof procedure for the logic of interest. This is supported by a user-friendly graphical interface. LoTREC should therefore ease both the task of students in learning and the task of researchers in debugging and prototyping.

LoTREC can either be run on-the-fly via the web, or be installed on the user's computer. It is implemented in Java and is therefore cross-platform: it can be run under operating systems such as Windows, Unix, Macintosh, and Solaris.

The implementation of LoTREC was done in the framework of several Master and PhD theses. It started with David Fauthoux's 2000 MSc thesis, that was based on ideas from a 1997 Fundamenta Informaticae paper by Marcos Castilho, Luis Fariñas del Cerro, Olivier Gasquet and Andreas Herzig [CFdCGH97]. Mohamad Sahade's 2006 and Bilal Said's 2009 PhD theses worked things out in detail, both in theory and implementation.

Overview of the Chapters

The first two chapters introduce Kripke models and the language of modal logic. The rest of the chapters present the tableaux method for various families of modal logics.

Chapter 1. Modelling with Graphs In the first chapter we introduce Kripke models. With the help of a series of examples we present the main modal concepts that can be interpreted by means of possible worlds semantics: the concepts of event, action, program, time, belief, knowledge, and obligation. In the end of the chapter we show how Kripke models can be built by means of LoTREC and give a formal definition of Kripke models.

Chapter 2. Talking About Graphs In the second chapter we introduce a general formal language in order to talk about properties of Kripke models. We then show how such languages can be defined in LoTREC and how to check that a formula is true in a given world of a given Kripke model. The latter is called model checking. Beyond model checking we also introduce the reasoning tasks of satisfiability checking, validity checking, and model building. We show that all other tasks can be reduced to the latter, which we focus on in the rest of the book: we show how to build models for a series of logics. These logics are grouped into families, according to the techniques the tableaux implementation in LoTREC requires.

Chapter 3. The Basics of the Model Construction Method This chapter is about the basic modal logic **K**and its multimodal version $\mathbf{K}_n$. We also present the basic description logic **ALC**, which can be viewed as a notational variant of $\mathbf{K}_n$. The implementation of reasoning methods for all these logics can be done by means of the most basic rules, that are combined by the most basic strategies: fair strategies.

Chapter 4. Logics with Simple Constraints on Models This chapter is about the model construction problem in some classes of models: models satisfying the conditions of reflexivity, seriality, symmetry, and combinations thereof. The corresponding logics are **KT**, **KD**, **KB**, **KTB**, and **KDB**. We also consider models whose accessibility relation is confluent (logic **K.2**) or is an equivalence relation (logic **KT45**, alias **S5**). While these conditions are only about a single relation, we also study properties involving two accessibility relations: inclusion and permutation. The reason for grouping these classes of models in a chapter is that model construction can be implemented by means of the most basic rules. We therefore call all these constraints *simple*. In the end of the chapter we present a general termination theorem that holds for almost all of these logics and which guarantees that the tableau construction does not loop.

Chapter 5. Logics with Transitive Accessibility Relations This chapter is about the model construction problem in classes of models satisfying the constraint of transitivity. We present the modal logics of the class of models where the accessibility relation is transitive (**K4**), transitive and serial (**KD4**), and transitive and reflexive (**KT4**, alias **S4**). For these logics, the model construction procedure may loop, which contrasts with the simple logics of Chap. 4. Termination can be ensured by means of blocking techniques: basically, the construction is stopped when the labels of a node are identical to those of some ancestor node. In the end of the chapter we present another general termination theorem guaranteeing that the tableau construction does not loop and which applies to all these logics.

Chapter 6. Model Checking This chapter shows how to implement model checking in LoTREC. There are two reasons why this topic is placed here: first, the implementation of model checking in LoTREC requires us to extend the tagging primitives of Chap. 5; second, model checking is going to be used in the next chapter.

Chapter 7. Modal Logics with Transitive Closure This chapter is about the model construction problem in classes of models having relations that are transitive closures of other relations. The main such logics are linear-time temporal logic **LTL** and propositional dynamic logic **PDL**. These logics require both blocking and model checking.

As we have already pointed out, this book not only presents the most important modal logics, but also tableaux proof procedures for them. We start by giving the LoTREC primitives for building models (Chap. 1) and for defining the language (Chap. 2). In an incremental way, each of the following chapters introduces the LoTREC primitives that are needed in order to automatically build models for the logics it is about.

Audience and Prerequisites

Readers of this book need only little background in mathematics and logic: some knowledge of the bases of propositional logic is sufficient. It is intended for undergraduate students ranging from the humanities and social sciences to computer science and mathematics.

Acknowledgments

Thanks are due to several persons who helped us to launch the idea and to improve this book. This is the place to warmly thank all of them.

The members of the Logic, Interaction, Language and Computation group (LILaC) at the Institut de Recherche en Informatique de Toulouse (IRIT) provided support and feedback throughout the years. David Fauthoux and Mohamad Sahade respectively wrote their Master and PhD thesis on LoTREC. Philippe Balbiani, Luis Fariñas del Cerro, Dominique Longin, and Emiliano Lorini gave us precious advice. Philippe Balbiani, Teddy Bouziat, Emiliano Lorini, and Faustine Maffre tested

LoTREC and read parts of the manuscript, allowing us to correct several mistakes. LILaC, IRIT and Toulouse are wonderful places to be and to work!

The material in the present book was presented partly or entirely in several workshops, conferences, and tutorials at spring and summer schools, in particular at the 2009 Universal Logic (UNILOG) spring school and at the 2009 European Summer School in Logic, Language and Information (ESSLLI 2009). The questions, remarks and comments of the attendees allowed us to improve the presentation.

Finally, many users of LoTREC—several of which used LoTREC when teaching introduction to modal logic classes—sent us encouraging feedback and suggested some improvements. To get an email saying that LoTREC was simple and easy to use was always a pleasure for us!

Toulouse, France
July 2013

Olivier Gasquet
Andreas Herzig
Bilal Said
François Schwarzentruber

Contents

Chapter 1
Modelling with Graphs

Science is basically about modelling various phenomena of the world. The concepts of space, time and events occur in most of these modellings. If moreover agents are part of the phenomenon to be modelled—be they human or artificial—then the concepts of action, knowledge, belief and goal play an important role. *Graphical models* are prominent examples of models for these and many other concepts. A graphical model is made up of *nodes* and *edges* between nodes; both nodes and edges may have *labels*.

Actually the term 'modelling' has two different understandings. First, it may be understood as representing a concrete part of the world: a particular device or a type thereof, a particular system, a particular situation, etc. For instance, we may model an agent's knowledge in a particular situation by a particular graph, as we shall do in Sect. 1.3 of the present chapter; or we may model the family relations within a given, fixed set of individuals, as we shall do in Sect. 1.5.

Second, modelling may be understood as representing more abstract properties of the world. In the latter case we also talk about theories that model some concept, such as theories of time or theories of knowledge. For instance, we model a theory of knowledge by a class of graphs, as we shall discuss in Sect. 1.9; or we may model various kinds of family relationships by a class of graphs where e.g. the grandfather relation is the composition of the 'parent' relation with the 'father' relation.

In the present chapter we start by exploring the first understanding of graphical models: we model a series of very simple examples—somewhat arbitrarily called 'situations,' 'systems,' or 'machines'—by means of graphs:

- the functioning of a light bulb according to what actions are performed on the toggle switch commanding it;
- the actions and events occurring when one locks or unlocks a car with a remote key;
- a traffic light switching its colour during consecutive time periods;
- the knowledge of players during a simple card game;
- a driver's obligations according to traffic regulations;
- the genealogical links between agents.

O. Gasquet et al., *Kripke's Worlds*, Studies in Universal Logic,
DOI 10.1007/978-3-7643-8504-0_1, © Springer Basel AG 2014

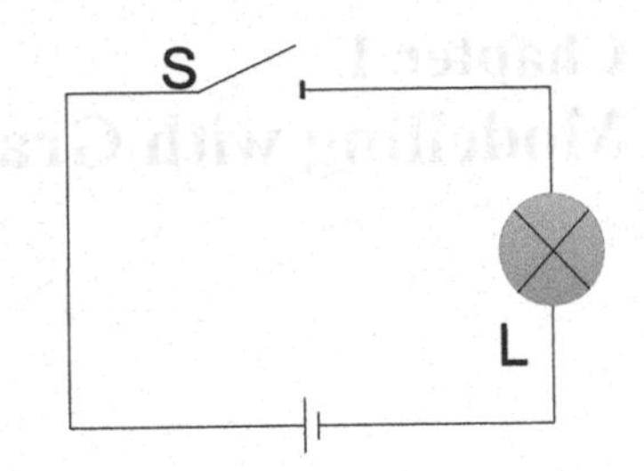

Fig. 1.1 A light and a switch

In all but the last of these examples graph nodes are viewed as *possible worlds*. They can be identified with complete descriptions of a possible state of the world or at least, complete descriptions of its relevant features. In contrast, one might also view nodes as *objects of the world* and edges as relations between these objects. This is the point of view taken in the last example and more generally in semantic web ontologies.

We then show how to construct the graphs modelling these examples with the aid of our tool LoTREC. This introduces the first part of LoTREC's declarative language: the language allowing us to construct and describe labelled finite graph structures.

We end the chapter by examining the second understanding of graphical models: we associate classes of graphs to theories of several important concepts, viz. to the concepts of time, action, programs, mental attitudes, and ontology.

1.1 Actions, Events, Programs

A Light and a Switch Let us start by modelling an electric circuit consisting of a light bulb and a switch as depicted in Fig. 1.1.

One way of talking about that circuit is: (1) "When the light is off, then toggling the switch turns the light on." This is certainly a true statement. But there is more to say, and we may start to think about other states of this circuit and say: (2) "When the light is on, then toggling the switch turns the light off." This is also true about our little circuit. One might describe it as well by other statements such as "The light can be either on or off," "Toggling is the only action we can perform," and so on. But let us focus on statements (1) and (2).

Suppose we want to restrict our view of the world to this circuit. In order to represent its functioning we may draw on a piece of paper the two possible states of the light: LightOn and LightOff, and we may represent the Toggle action by means of two *edges* linking these states and labelled by the name of the action Toggle. This is depicted in Fig. 1.2.[1]

[1] All the graphics of the form of Fig. 1.2 (nodes as blue rectangles and edges as blue arrows) were generated automatically with LoTREC. We only arranged some of them a bit by drag-and-drop rearranging the nodes in order to make the display more symmetric. Sometimes we also added a name to the nodes in order to be able to refer to them in the text.

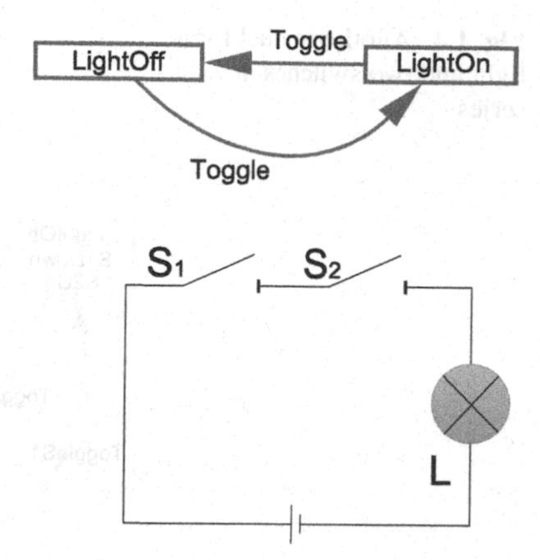

Fig. 1.2 The two possible states of the light

Fig. 1.3 Two switches in series

One might wonder why in the example above we have two labels LightOn and LightOff: one may only use one of them, say LightOn, and stipulate that the absence of LightOn indicates that LightOn is not the case.

A Light with Two Switches in Series Let us try to model the electric circuit of Fig. 1.3. Just as we did for the first example, we should start by describing the functioning of this circuit. Then we should try to extract a graph that represents it as rigorously as possible.

At first sight, we notice that the circuit consists of a light and two switches in series and that the light goes on when both switches S_1 and S_2 are toggled downward. Otherwise, if one of the switches remains in the upper position then the light stays off. Let us suppose that only one action can be executed at a time, i.e., it never happens that both switches are toggled simultaneously. First, in addition to the state of the light (on or off), the states should contain information about the state of each switch (up or down). The set of possible states can then be identified by the following four configurations of labels:

1. LightOn, S1Down and S2Down;
2. LightOff, S1Up and S2Down;
3. LightOff, S1Down and S2Up;
4. LightOff, S1Up and S2Up.

Second, a transition between two of these states should be labelled by the action that is applied on one of the switches. For example, if the circuit is in state 1, then toggling S_1 alters the state of the circuit to state 2. The resulting model is shown in Fig. 1.4.

Exercise 1 Describe the functioning of the circuit of Fig. 1.5, and draw the graph modelling it.

The Self-locking Car Let us consider a car that its owner can lock/unlock by clicking her remote key. We suppose that the car has an auto-lock facility which locks the car in case its owner forgot to do so.

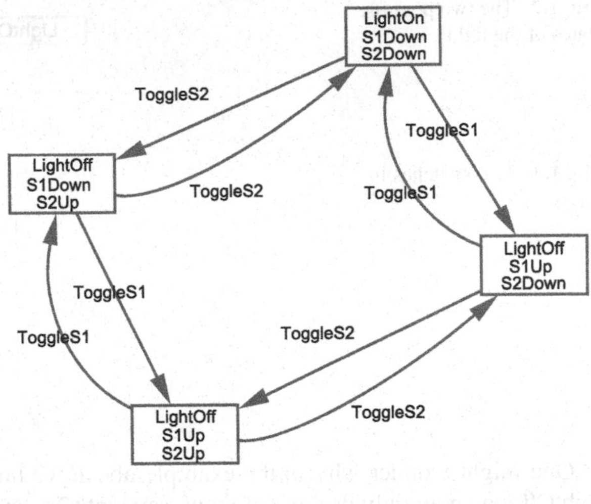

Fig. 1.4 Another model for a light and two switches in series

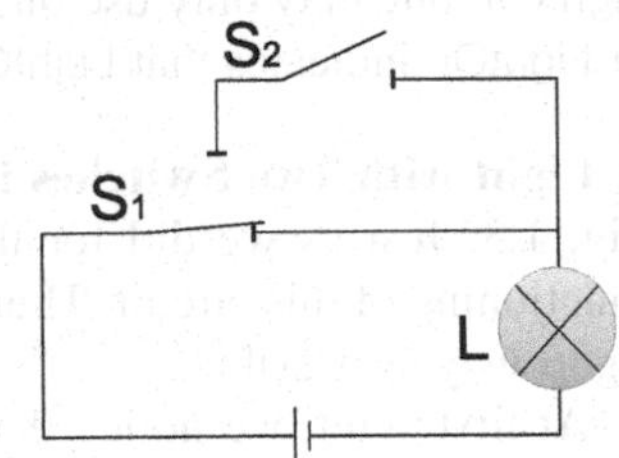

Fig. 1.5 Pearl's circuit [Pea00]

At any moment, the user may perform one among six actions:

- Click: the effect is to lock or unlock the door according to whether it was locked or not just before. We suppose that it cannot be performed when the key is inside the car, nor when the door is open.
- Close: the effect is that the door of the car is closed. It can only be executed when the door is open.
- Open: the effect is that the door of the car is open. It can only be executed when the door is closed.
- Wait: has no immediate effect, but causes the door to be locked after a short interval due to the auto-lock facility. The action Wait can always be executed. If the door is already locked it does not change anything; so we may consider that in this case, the transition is from one state to itself.
- GetIn and GetOut: the actions, for the owner, of getting in or out of the car. They can only be executed when the door is open and when the user is outside in the case of the former action, and inside in the case of the latter.
- PutKeyIn and PutKeyOut: the effects and conditions are similar to that of GetIn and GetOut, but concern the key of the car.

The relevant part of the functioning of the car is described in Fig. 1.6. The node labels are self-explanatory: KeyIn, KeyOut, UserIn, UserOut, Locked, Unlocked, Closed and Open.

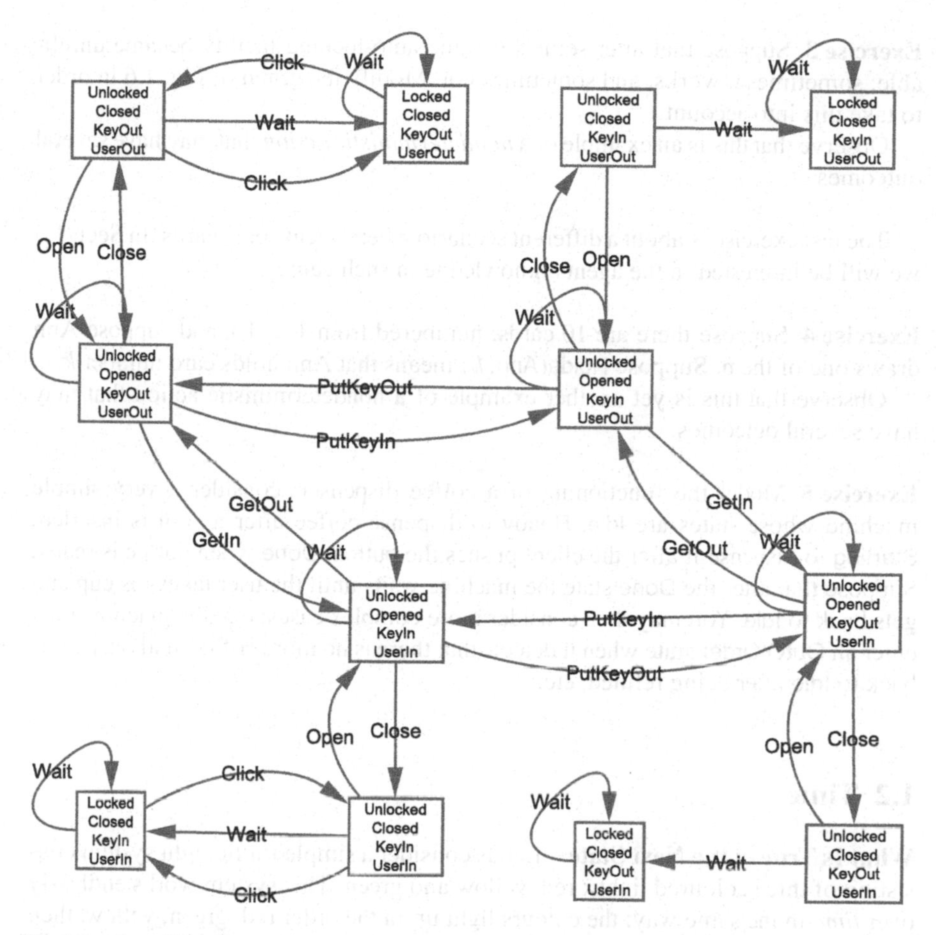

Fig. 1.6 The self-locking car

The following exercises amount to doing by hand what one can do with a computer in order to check whether our car locking system is safe.

Exercise 2 Provide evidence that the system of Fig. 1.6 is not safe: prove that it may happen that the owner is outside the car, the key is inside, and the auto-locking facility locks the key up.

Do so with the help of the graph of Fig. 1.6. Suppose that initially the user and the key are outside, and find a sequence of actions such that afterwards:

- The key is locked in the car, the owner being outside (so she will need her spare key... if she has it with her);
- The owner is locked in the car, the key being outside (hopefully she has a spare key or a hammer with her).

Exercise 3 Suppose that after some years the auto-locking facility became unreliable: sometimes it works, and sometimes not. Modify the graph of Fig. 1.6 in order to take this into account.

Observe that this is an example of a *nondeterministic action* that may have several outcomes.

The last exercise is about a different scenario where agents play cards. In Sect. 1.3 we will be interested in the agents' knowledge in such games.

Exercise 4 Suppose there are 10 cards, numbered from 1 to 10, and suppose Ann draws one of them. Suppose Holds(Ann, k) means that Ann holds card number k.

Observe that this is yet another example of a nondeterministic action that may have several outcomes.

Exercise 5 Model the functioning of a coffee dispenser: consider a very simple machine whose states are Idle, Ready to dispense coffee after a coin is inserted, Starting to dispense it after the client pushes the button, Done when coffee is ready. Suppose that after the Done state the machine waits until the user takes his cup and gets back to Idle. You may then consider more complex cases: e.g. the machine may enter an OutOfOrder state when it detects that there is no more coffee, and only turns back to Idle after being refilled, etc.

1.2 Time

What Is True in the Next State Let us consider a simple traffic light system consisting of three coloured lights: red, yellow and green. This system works endlessly over *time* in the same way: the colours light up in the order red, green, yellow, then red again, and so on.

The states of the system and the relation between them are easy to extract from the above description. A graph representing a traffic light can be drawn with three states: Red, Yellow and Green, and the relation between these states reckons for the order between these states.

The notion of time appears clearly in the periodic succession of these states: we chose to link the state Red to Green, and we chose to label the edge linking them by Next to express the order between the moments of their appearance. Similarly, we link Green to Yellow and Yellow to Red by edges labelled Next. The resulting graph is illustrated in Fig. 1.7.

Exercise 6 The above is a French traffic light. Suppose now that the sequence is not simply 'red, green, yellow' but—as in may countries—'red, red and yellow, green, yellow, red.' Modify the graph in Fig. 1.7 in order to model that. Do the same for the sequence red, yellow, green, yellow, red.

Can you draw these models with only two nodes? Why would the resulting behaviour of the traffic light be undesirable in the second case?

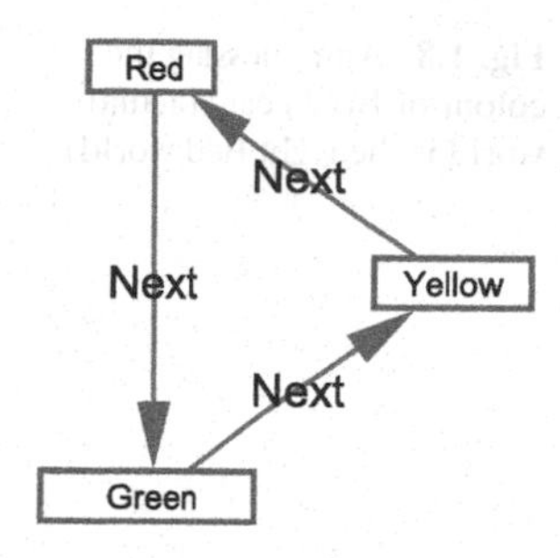

Fig. 1.7 Traffic lights model

What Is True in Some Future State Suppose we are not interested in the 'next state' relation, but in the 'eventually' relation between nodes, i.e., in what may become true in some state that is in the future of the current state. Add this relation to Fig. 1.7.

That relation is clearly a transitive relation, and—if we consider a non-strict sense of 'eventually'—it is also a reflexive relation.[2] Actually the 'eventually' relation is the *transitive closure* of the 'next state' relation, i.e. the relation obtained by adding all and only those edges needed to make the relation transitive.[3]

Future and Past Instead of the future we may be interested in the past, and focus on the *converse* of the 'next' relation.[4] Add the converse of the 'next' relation to Fig. 1.7.

Exercise 7 Model the four seasons, where the 'next' relation is the transition from one season to another.

1.3 Knowledge and Belief

Knowing the Colour of the Card Consider a very simple card game with just two cards. One of them is red and the other one black. The cards are lying face down on the table, and Ann and Bob cannot see their colour (but they have been told before that there is a red and a black card). Bob then picks up a card and watches it; let us say it is red. Ann cannot see which colour it is, and both know that. Ann has to guess the colour of the card.

Suppose that we are interested in answering the following questions: what will Ann's answer be? Can she be sure of it? Does Bob know this? Is Bob able to say if Ann's answer is correct?

---

[2] A relation R is transitive if and only if for every x, y, z we have that if $\langle x, y\rangle \in R$ and $\langle y, z\rangle \in R$ then $\langle x, z\rangle \in R$. A relation R is reflexive if and only if for every x we have that $\langle x, x\rangle \in R$.

[3] More formally speaking, the transitive closure of a relation R is the smallest transitive relation containing R.

[4] The converse of a relation is obtained by inverting the sense of the edges. Formally, the converse R^{-1} of a relation R is defined as: $R^{-1} = \{\langle y, x\rangle \mid \langle x, y\rangle \in R\}$.

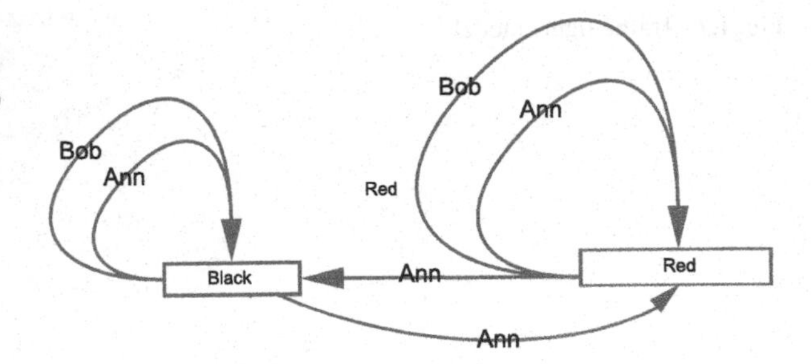

Fig. 1.8 Ann guessing the colour of Bob's card (actual world is the right Red world)

To answer these questions we have to figure out first what Ann knows about the card and what Bob knows about it. Other interesting questions are whether Ann knows what Bob knows and whether Bob knows what Ann knows. In other words, we are interested in modelling the *knowledge* of Ann and Bob. That knowledge might be *higher order*: knowledge about the knowledge of another agent, or even about what that other agent knows about a third agent, etc.; and similarly for belief.

In our scenario Bob knows that the chosen card is red, while Ann does not. However, Ann knows that it is either black or red, and she knows that Bob knows the colour of the card. Moreover, being as intelligent as us, Bob knows all we have said above.

In order to give a graphical representation of these statements we will proceed exactly as in Sect. 1.1: we will try to figure out (1) what the possible states are, and (2) what the relations between these state are.

Concerning the colour of the selected card there are two possibilities: either it is red—abbreviated as Red—, or it is black—abbreviated as Black. Thus we draw two nodes with these labels. The Red state is the actual state of the world.

It remains to add to our graph the representation of Ann's and Bob's knowledge about these states. We can use the edges to this end, just as we did in Sect. 1.1 to depict the actions that allow one to go from a state to another. The difference is that here we have to represent which states are *envisioned* by Ann and by Bob, or *indistinguishable* for Ann and for Bob. For instance, when the actual state is Red, Ann still considers that Black is a *possible* state: she does not have enough information to distinguish these two states. Thus we link Red to Black with an edge labelled by her name Ann. We also link Black to Red with an edge labelled Ann since Red is a possible state (where Ann thinks that the card is actually Black). This does not hold for Bob, since there is no uncertainty for him about the colour. So whenever he thinks that it is red then the state Black is not possible for him; and vice-versa. Coming back to Ann, when she thinks that the actual state is Red (resp. Black) then she is also considering Red (resp. Black) as a possible state. Thus we link each state to itself with an edge labelled Ann. The same reasoning can be made about Bob's knowledge. That is why we also have reflexive edges labelled Bob.

This modelling of Ann's and Bob's *epistemic states* is depicted in the graph of Fig. 1.8.

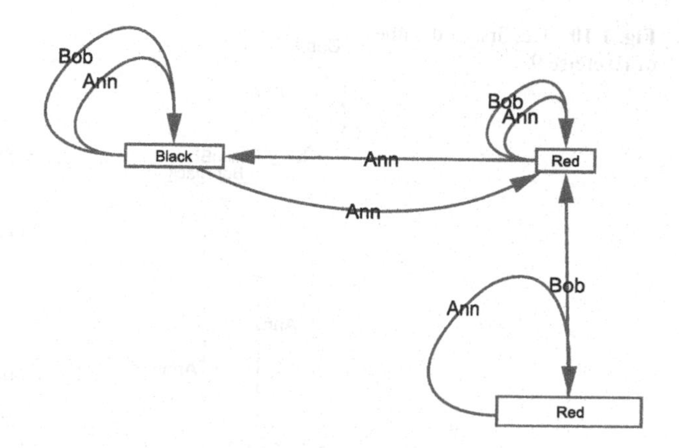

Fig. 1.9 Ann is cheating (actual world is the bottom Red world)

Knowledge vs. Belief: What if Ann Is Cheating? Let us now modify our scenario a bit, and suppose that Ann *can* see the colour of the card Bob selects (say because there is a mirror behind him) and that Bob does not know this.

Suppose that Bob picked up the same red card again. What should the model of Fig. 1.8 look like now? Let us try to figure it out together.

In fact, as for Bob, the situation has not changed at all: he still thinks that Ann does not know the colour of the selected card, that she hesitates whether it is red or black, and that he is the only one who knows that it is red. As for Ann, she now knows what is the actual world: she is sure that the card colour is red and she knows that Bob thinks that she does not know what the actual world is.

To describe this graphically we have to change the model of Fig. 1.8 a bit. We reuse its entire graph to represent Bob's mental state. However, we no longer consider its Red state to be the actual world.

On the other hand, in order to represent Ann's mental state we draw yet another state labelled by Red. As this is the only possible state for Ann, we link it to itself by an arrow labelled Ann. Finally, in order to represent that she knows what Bob is thinking, we link this state to the Red state of the graph part that we copied from Fig. 1.8.

The resulting model should look like the graph of Fig. 1.9. It is "Ann's Red state" which is the actual world now. Note that "Bob's Red state" has exactly the same labels: they do not differ in the facts, but only in the agents' beliefs.

Exercise 8 Suppose now that Ann and Bob are playing with a third person called Cathy, and suppose that when Bob chooses the card he hands it over to Cathy without watching it. Change the graph of Fig. 1.8 in order to reflect this.

Exercise 9 Ann and Bob are playing another game now. Each of them chooses a card from a stack of four cards that are either red or black. Suppose that Fig. 1.10 is modelling correctly the mental state of Ann and Bob after they have chosen their cards. The actual world is labelled AnnRed and BobRed.

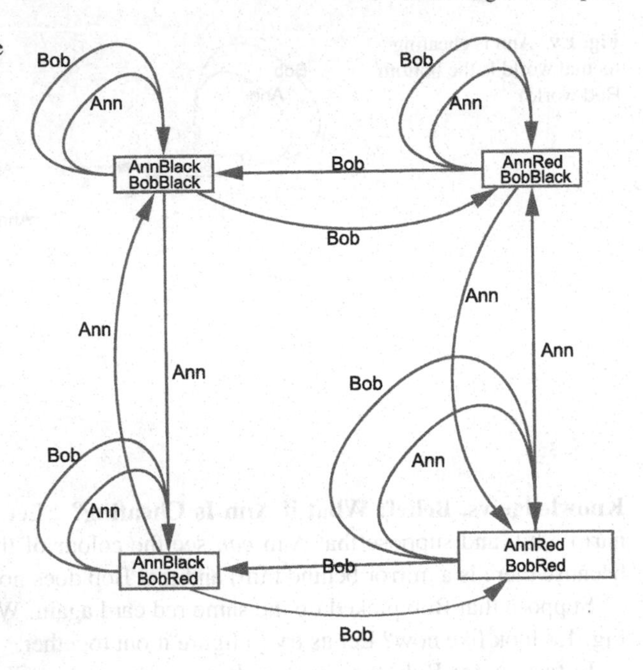

Fig. 1.10 The first card game of Exercise 9

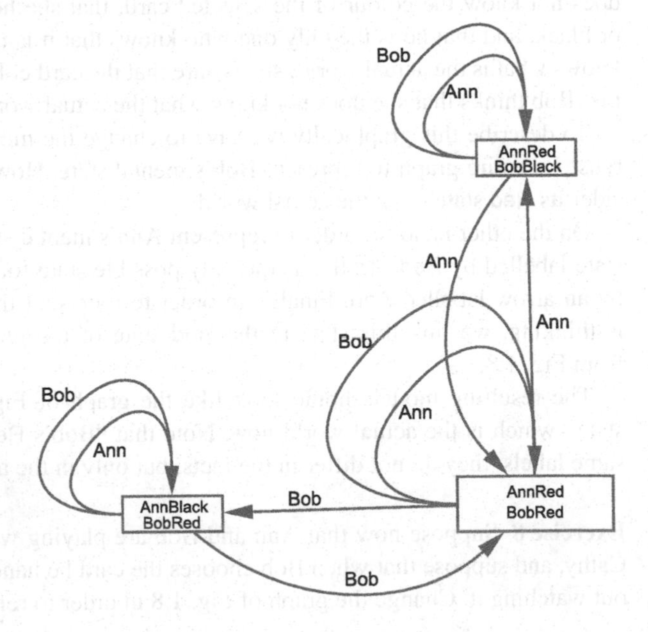

Fig. 1.11 The second card game of Exercise 9

1. How many red cards and how many black cards were there in the stack (according to Ann and Bob's beliefs)?
2. What if the correct model was the one in Fig. 1.11?

Fig. 1.12 Four prisoners wearing four black hats

Exercise 10 (Muddy children) Two children called I_1 and I_2 come back from the garden after having played for the whole afternoon. Both have dirty foreheads. However, each of them can only see the other's forehead, but not his own. Therefore each child knows that the other child is dirty, but does not know whether he himself is dirty or not.

1. Draw a Kripke model of that situation.
2. Extend the previous scenario to 3 children I_1, I_2, I_3. Draw a Kripke model of that situation.
3. Generalize to n children $I_1, \ldots, I_n$ whose foreheads are all dirty. Every child can see each other's foreheads and therefore knows that all others are dirty, but does not know whether he is dirty or not.

Exercise 11 (Prisoners' hats) Model the following situation. There are n prisoners $I_1, \ldots, I_n$ each of whom wears a hat. The hats are either black or white, and each of the prisoners knows that. They are lined up: I_1 can see the colours of all other prisoners except his own; I_2 can see the colours of all other prisoners except those of I_1 and I_2; etc. Generally, I_k can see the colours of all prisoners except those of $I_1, \ldots, I_k$. The scenario is depicted in Fig. 1.12.

1. Draw a graph of the situation for $n = 1$ (which makes sense!), 2, 3, 4.
2. Now suppose each prisoner has to make a statement about the colour of his hat. When all prisoners have spoken then they are all freed if there is at most one wrong statement; and they are all executed if more than one of them made a mistake. For the cases $n = 2, 3, 4$, draw the graph after the first prisoner has announced his colour.
3. Suppose the prisoner are allowed to discuss and settle a strategy before getting their hats. Find a sequence of statements of each agent starting with I_1 which guarantees that all of them are freed.

Remark 1 Would there be something wrong with a node labelled both AnnRed and AnnBlack? Logically speaking, nothing: a graph with such a node is a legal graph! Indeed, nothing *logically* prevents a card from being both red and black. If one wants to exclude such states one may only make use of one of these two labels (say, AnnRed) and consider that if a card is not red then it is black. (Being black is therefore identified with being not-red.)

We have modelled each agent's knowledge in our examples by means of relation of indistinguishability. Anticipating a bit, we observe that every such relation is naturally an *equivalence relation*: a relation that is *reflexive*, *reflexivity transitive*, and *symmetric*. 'Reflexive' means that each node is indistinguishable from itself; 'transitive' means that if w_1 is indistinguishable from w_2 and w_2 is indistinguishable from w_3 then w_1 is indistinguishable from w_3; and symmetric means that if w_1 is indistinguishable from w_2 then w_2 is indistinguishable from w_1. The property of symmetry can be replaced by *euclideanity*: if w is indistinguishable from w_1 and w is indistinguishable from w_2 then w_1 is indistinguishable from w_2. We have also seen that the relation for belief is not necessarily reflexive, see Bob's relation in Fig. 1.9. It is not symmetric either. It is however serial (also called idealisable): for every w there is at least one w' that is indistinguishable from w; moreover it is transitive and Euclidean.

Exercise 12 Find a graphical model of the agents' knowledge in the following situation: an ex-couple meets and each starts to wonder if the other is in a relationship. (Hint: being or not being in a relationship might to some extent be compared to having a red or a black card.)

1.4 Obligations and Permissions

Let us take up the traffic light example as depicted in Fig. 1.7. Consider the state where the traffic light is red and suppose that Ann doesn't stop at the red light. How can we model that Ann is violating the traffic regulations?

In this section, we explain that we can model obligations and permissions with possible worlds. In Sect. 1.3, the graph nodes were viewed as being either the actual world or worlds that are compatible with an agent's knowledge. Here we view nodes as permitted worlds, i.e., worlds that are compatible with the agents' obligations. In order to completely describe a situation, in the actual world we have to consider all the permitted worlds. More precisely, we link the actual world to each permitted world. A permitted world is often called an *ideal counterpart* of the actual world.

In our example depicted in Fig. 1.13, there is only one ideal counterpart world of the world that is labelled Red and AnnDrives: it is labelled by AnnStops and Red. Then we may say that Ann is *obliged* to stop because all ideal counterparts of the Red-AnnDrives world (there is only one!) are labelled AnnStops. We may also say that Ann is *forbidden* to drive because no ideal counterpart of the Red-AnnDrives world is labelled AnnDrives.

Now consider a world that is labelled Green and AnnDrives, and that is not labelled AnnStops. That world is an ideal counterpart of itself because the actual world is permitted. It is therefore not the case that all its ideal counterparts are labelled by AnnDrives: indeed Ann is also permitted to stop. The Green-AnnDrives world also has an ideal counterpart that is labelled Green and AnnStops and that is not labelled AnnDrives; therefore Ann is not obliged to drive either: she is permitted to stop because there is at least one ideal counterpart that is labelled AnnStops. Figure 1.14 shows the corresponding graph.

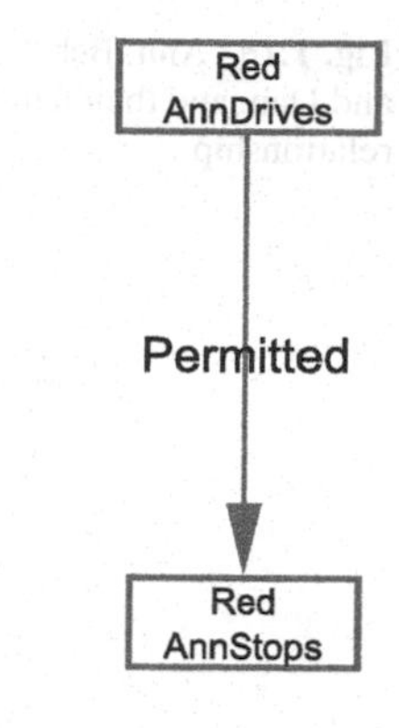

Fig. 1.13 Ann has to stop

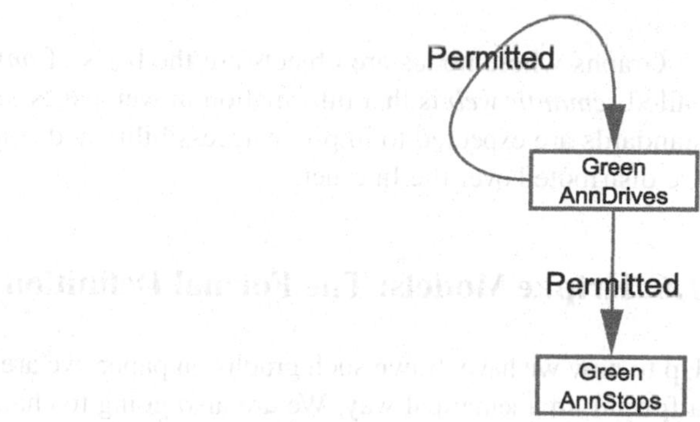

Fig. 1.14 Ann is permitted to drive, as well as to stop

Consider a possible world having no ideal counterparts at all. It means that in that word nothing is permitted. Such a situation is undesirable for a system of regulations and norms! In other words, every possible world should have at least one ideal counterpart. This constraint is called *idealisation* or *seriality*.

1.5 Relations Between Objects

Up to now we have viewed graph nodes as possible worlds, also called states of affairs. But they may as well be viewed as objects of some domain: humans, physical objects, companies, and the like. These objects are related by relations. Let us only consider binary relations: relations with two object arguments.

Suppose Ann and Bob are siblings, Ann is the mother of Cathy, and Cathy is the mother of Dan. Ann and Cathy are female, while Bob and Dan are male. The relationship between these persons is depicted in Fig. 1.15.

Exercise 13 Beyond the unary relations Male and Female labelling the nodes and the binary relations Sibling and MotherOf labelling the edges, one may extract other information from that graph, such as about the unary relation isMother and the binary relations GrandmotherOf, BrotherOf, SisterOf, and isUncleOf. Check that Ann is a grandmother of Dan and a sister of Bob. Check that all of Cathy's children are male.

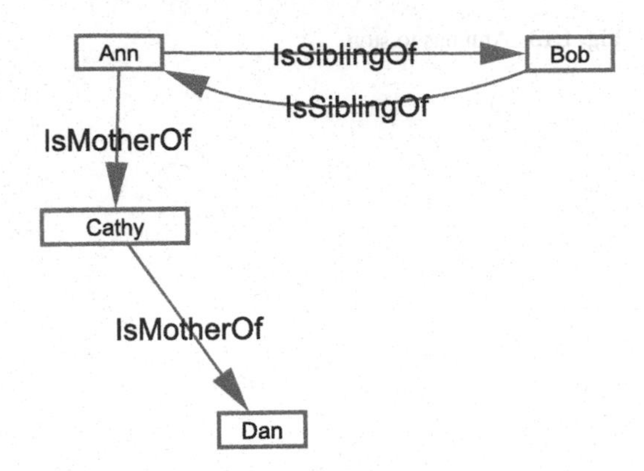

Fig. 1.15 Ann, Bob, Cathy, and Dan, and their family relationship

Graphs whose nodes are objects are the basis of *ontologies*. The idea of the so-called *semantic web* is that information in web pages refer to such ontologies. Such standards are expected to improve accessibility and consistency of information that are distributed over the Internet.

1.6 Kripke Models: The Formal Definition

Up to now we have drawn such graphs on paper: we are now going to define them in a formal, mathematical way. We are also going to change our terminology in order to conform to the literature on modal logic: graphs are called *Kripke models* after Saul Kripke who was the first to introduce and study them in a systematic way as a semantics for modal logics; graph nodes are called *possible worlds* or *states*; the set of edges decorated by a given edge label make up the *accessibility relation* for that label; the set of possible worlds decorated by a given node label make up the *valuation* of that label. Finally, graphs without node labels are called *frames*. The latter are important for technical reasons, as we shall see later on.

Definition 1 (Kripke frame, Kripke model) Given a countable set of edge labels $\mathcal{I}$, a Kripke frame is a tuple $\langle W, R\rangle$ where:

- W is a non-empty set, called the set of possible worlds;
- $R: \mathcal{I} \longrightarrow 2^{W\times W}$ maps each $I \in \mathcal{I}$ to a binary relation $R(I)$ on W, called the accessibility relation for I;

Given a countable set of atomic node labels $\mathcal{P}$ and a countable set of edge labels $\mathcal{I}$, a Kripke model is a tuple $M = (W, R, V)$ where:

- $\langle W, R\rangle$ is a Kripke frame;
- $V: \mathcal{P} \longrightarrow 2^{W}$ maps each $P \in \mathcal{P}$ to a subset of W, called the valuation of P.

We say that the model M is a model over the frame $\langle W, R\rangle$.

For $M = (W, R, V)$, every couple $\langle M, w\rangle$ such that $w \in W$ is called a *pointed model*.

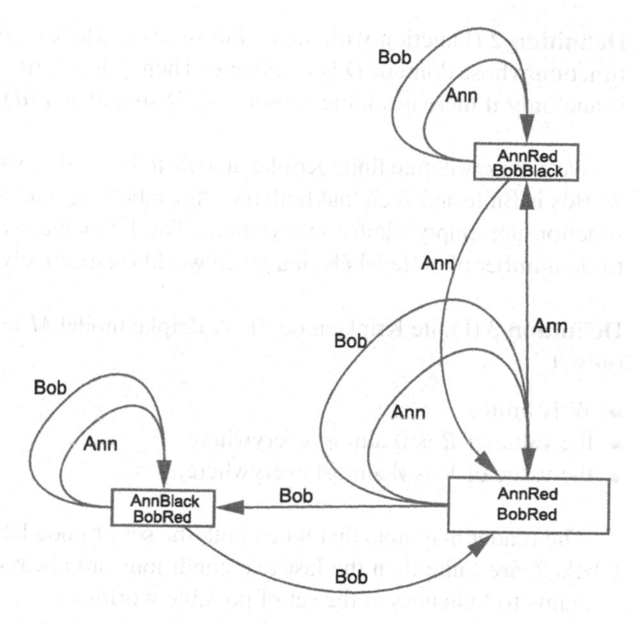

Fig. 1.16 Ann and Bob playing cards (actual world is the world where AnnRed and BobRed are true)

The functions R and V can be understood as giving meaning to ('interpreting') the node and edge labels of the language. If we think of the elements of $\mathcal{P}$ as propositions then we say "P is true at w" when $w \in V(P)$.

Example 1 Let us revisit the model for the card game as described by Fig. 1.11, that we repeat in Fig. 1.16 for convenience. It has three possible worlds, a set of edge labels $\mathcal{I} = \{\mathsf{Bob}, \mathsf{Ann}\}$, and a set of node labels $\mathcal{P} = \{\mathsf{AnnRed}, \mathsf{BobBlack}\}$. Let us suppose that the possible world at the top of the figure is called w_0 and the world just below w_0 is called w_1. Then w_0 belongs to both $V(\mathsf{AnnRed})$ and $V(\mathsf{BobBlack})$: AnnRed and BobBlack are the atomic propositions that are true at w_0; w_1 belongs to both $V(\mathsf{AnnRed})$ and $V(\mathsf{BobRed})$: AnnRed and BobRed are true at w_1. Worlds w_0 and w_1 are linked by an edge labelled Ann, but there is no link between them that is labelled Bob: $(w_0, w_1) \in R(\mathsf{Ann})$ but $(w_0, w_1) \notin R(\mathsf{Bob})$. Of course, worlds may be related to themselves; for example, $(w_0, w_0) \in R(\mathsf{Ann})$ and $(w_0, w_0) \in R(\mathsf{Bob})$.

Among the set of all Kripke models, the set of *finite* models is of particular interest: they can be immediately drawn on paper (at least in principle). In contrast, we can draw infinite models only if we resort to conventions such as the use of dots to represent infinitely many worlds. For example, the model of Fig. 1.16 is finite.

A finite Kripke model should not only have a finite set of possible worlds: we also have to exclude the case where infinitely many labels decorate some possible world or some edge. Let us formulate that in terms of our formal description of Kripke models of Definition 1. We need the following notion: we say that a function f with a countable domain D has some value *almost everywhere* if $f(d)$ equals that value for all but a finite number of elements d of D.

Definition 2 (Function with value almost everywhere) Suppose $f : D \longrightarrow S$ is a function whose domain D is countable. Then f has value $v \in S$ almost everywhere if and only if there is a finite subset $F \subseteq D$ such that $f(d) = v$ for every $d \notin F$.

We can now define finite Kripke models to be models such that the set of possible worlds is finite and such that both the edge labelling function and the node labelling function are empty almost everywhere. The latter means that there can only be a finite number of node labels in a given world (respectively: on a given edge).

Definition 3 (Finite Kripke model) A Kripke model $M = (W, R, V)$ is finite if and only if

- W is finite;
- the value of R is $\emptyset$ almost everywhere;
- the value of V is $\emptyset$ almost everywhere.

The reader may note that when both the set of node labels $\mathcal{P}$ and the set of edge labels $\mathcal{I}$ are finite then the last two conditions are always satisfied: then finiteness amounts to finiteness of the set of possible worlds.

Remark 2 An equivalent way to formulate models is to define the accessibility relation as a mapping from $W \times W$ to the powerset of $\mathcal{I}$, formally: $R : W \times W \longrightarrow 2^{\mathcal{I}}$. Similarly, one may define the valuation as a mapping from W to the powerset of $\mathcal{P}$, formally: $V : W \longrightarrow 2^{\mathcal{P}}$. Under such formulations, finiteness of the model $M = (W, R, V)$ can be defined as finiteness of W, of every set $R(w, u)$ and of every set $V(w)$.

1.7 Kripke Models: Building Them with LoTREC

LoTREC offers a language that allows us both to build graphs and to talk about graphs. We now introduce LoTREC's language for building graphs; the keywords for *describing* graphs will be introduced in the next chapter. You may read this section while simultaneously working with LoTREC. We recall that LoTREC is accessible at

http://www.irit.fr/Lotrec

You may in particular consult the video tutorials there.

Graphs are built by means of a set of *actions*. These actions take the following form:

- `createNewNode w`
- `link w u Label`
- `add w Label`

where `w` and `u` are nodes and `Label` is a label.

The action createNewNode w creates a new node in the graph and calls it w. The action link w u Label supposes the existence of two nodes w and u and creates an edge from w to u that is labelled by Label. The action add w Label supposes the existence of a node called w and adds Label to its labels.

In the above list of actions Label is a *constant*,[5] whereas w and u are *variables*. LoTREC substitutes the variables w and u by names of nodes when applying the rules containing them: when the action createNewNode w is performed on some graph then LoTREC creates a new node and assigns it to w; and when add w Label is executed then LoTREC adds the label Label to the node assigned to w.

All the graphs that we have seen up to now were produced by means of the above three LoTREC actions. For example, the graph of Fig. 1.2 was built by executing the following sequence of actions:

```
createNewNode w
createNewNode u
add w LightOn
add u LightOff
link w u Toggle
link u w Toggle
```

These actions are implemented in LoTREC by means of what is called a *tableau rule*, to be introduced in more detail in Chap. 3. This requires us to prefix the above by the following two lines:

```
Rule BuildExampleGraph
  isNewNode w
```

Moreover, the first action createNewNode w has to be dropped: LoTREC always starts by creating a root node anyway. That root node substitutes the variable w in isNewNode w when BuildExampleGraph is executed.

In practice, the rule BuildExampleGraph can be entered in the LoTREC interface as follows: start with a new logic; in the left "Loaded Logics" panel, select the rule tab; click on the "Add" button, enter the rule name (which is NewRule by default), then enter the condition isNewNode w in the "Conditions" panel (click on the "Add" button and select isNewNode in the drop-down list) and enter the actions one after the other in the "Actions" panel (click on the "Add" button and select the appropriate action).

It remains to call the rule BuildExampleGraph by a strategy in order for LoTREC to run. Strategies are defined in the strategy tab: open the default strategy (which is initially empty) by clicking on the "Edit" button and enter the rule in the "Code" panel. (You can avoid typing the rule name by clicking on the rule names

[5]By convention, expressions starting with a capital letter are *constants*. This is explained in detail in Sect. 2.2.2.

in the "Available Rules" panel.) You may also give a specific name to the strategy by changing its name from `DefaultStrategy` to e.g. `BuildMyModel`. In the case of our example the result is simply:

```
Strategy BuildMyModel
  BuildExampleGraph
```

This strategy only contains one rule. In general, strategies are more complex, as we shall see in the other chapters.

Graphs can be created, displayed, saved and loaded in LoTREC. More graph building primitives—those allowing us to tag nodes and formulas—are going to be introduced in Sect. 3.10.2 and in later chapters, but the reader may already discover them by playing with the LoTREC interface.

Exercise 14 Identify the actions required to build the graphs of Figs. 1.7 and 1.8 and write them down in the language of LoTREC.

Exercise 15 Write in the language of LoTREC a set of actions building the graph of Example 1.16.

You may consider saving these graphs built in the last exercises: we are going to use them in the next chapter in order to perform model checking. (Just click on the `SAVE` button to do so.)

1.8 The Typewriter Font Convention

In the previous section we have written node names such as `w` in typewriter font. Throughout the book we usc the following convention in order to highlight things that can be directly written in LoTREC:

- We use typewriter font when we are talking about things that are supposed to be understood by LoTREC and that can be implemented in LoTREC;
- We use standard math mode (italicised) when we write on paper things such as possible worlds, accessibility relations, valuations of Kripke models.

For example the node label P is understood as written on paper, while `P` is understood as represented in LoTREC; similarly for `M` and M, and for `w` and w; the graph building actions of Sect. 1.7 and the rule name such as `BuildExampleGraph` are written in typewriter font, while the definition of Kripke models of Sect. 1.6 is written in math mode.

As the reader may have noticed, beyond being in typewriter font, LoTREC keywords such as `createNewNode`, `link`, and `add` are underlined in the text.

1.9 Modelling Theories by Classes of Kripke Models

Up to now we have modelled concrete, selected parts of our surrounding world: each of our toy examples was represented by a single graph. The edges of that graph are modelling concepts such as time, programs, knowledge, etc. in that particular part of the world. For example, in the case of the cards game of Sect. 1.3 there is an edge from the Red node to the Black node that is labelled by the agent name Ann if and only if Ann is not able to distinguish these two nodes, given the knowledge she has in the Red node.

Suppose that we are not interested in Ann's knowledge in that particular situation but in the concept of knowledge itself. In each of our examples we have modelled an agent's knowledge by means of a relation of indistinguishability between states. As we have observed in the end of Sect. 1.3, relations of indistinguishability are always equivalence relations, i.e., they are reflexive, transitive, and symmetric. It is therefore a natural step to generally identify "Agent I knows that A" with "All states I cannot distinguish from the actual state are labelled by A," where A is *any* label. This amounts to viewing the class of graphs with equivalence relations labelled by agent names as *models of the theory of knowledge*. The theory of knowledge is therefore not modelled by a single graph but by a *class of graphs*. Such a theory of knowledge may not only be applied to human agents but also to artificial agents. Logics of knowledge are actually widely applied in distributed systems [FHMV95] and in multiagent systems [Woo02].

We may do the same for other concepts, such as time, events, actions, beliefs, goals, intentions, obligations, and so on: we may identify particular classes of models as representations of a theory of the concept under study. Let us summarise the constraints that are natural for the respective concepts and that we have already mentioned in Sects. 1.2 to 1.5.

- Time: if we suppose that the flow of time is linear then the 'next' relation should be deterministic, i.e., if w is related to both w_1 and w_2 then w_1 and w_2 are identical; and the 'henceforth' relation should be the reflexive and transitive closure of the 'next' relation.
- Events, programs, and actions: no particular properties.
- Knowledge: for every agent I, the relation associated to I should be an equivalence relation (reflexive, transitive and Euclidean).
- Belief: for every agent I, the relation associated to I should be serial (alias idealisable), transitive, and Euclidean. These constraints are weaker than those for knowledge because they do not require reflexivity.
- Obligation: for every agent I, the relation associated to I should be serial (alias idealisable).

We summarise the most important properties of a single binary relation R in Table 1.1. They are most prominent in the modal logic literature and were intensely discussed and studied. For each property we give its denomination in mathematics, the standard name of that constraint in Kripke models, and its definition.

One may identify particular subsets of the class of all Kripke models by means of constraints on the relation between nodes, on the node labels, or on the basic

Table 1.1 Some properties of accessibility relations

Property	Definition
reflexivity	for every w, $\langle w, w\rangle \in R$
seriality	for every w there is a w' such that $\langle w, w'\rangle \in R$
determinism	for every w, if $\langle w, w_1\rangle \in R$ and $\langle w, w_2\rangle \in R$ then $w_1 = w_2$
symmetry	for every w_1, w_2, if $\langle w_1, w_2\rangle \in R$ then $\langle w_2, w_1\rangle \in R$
transitivity	for every w_1, w_2, w_3, if $\langle w_1, w_2\rangle \in R$ and $\langle w_2, w_3\rangle \in R$ then $\langle w_1, w_3\rangle \in R$
euclideanity	for every w, w_1, w_2, if $\langle w, w_1\rangle \in R$ and $\langle w, w_2\rangle \in R$ then $\langle w_1, w_2\rangle \in R$
confluence	for every w, w_1, w_2, if $\langle w, w_1\rangle \in R$ and $\langle w, w_2\rangle \in R$ then there is u such that $\langle w_1, u\rangle \in R$ and $\langle w_2, u\rangle \in R$
(right) linearity	for every w, w_1, w_2, if $\langle w, w_1\rangle \in R$ and $\langle w, w_2\rangle \in R$ then $\langle w_1, w_2\rangle \in R$ or $\langle w_2, w_1\rangle \in R$

sets of labels `L` of the underlying language. For example, we may require that all relations are *reflexive*, *symmetric* or *transitive*; or we may require that some special labels—called names for worlds, or *nominals*—decorate a *unique* node.

The models satisfying a given set of constraints make up a *class* of models.[6] Through Chaps. 4 to 7 we shall meet many of these classes.

1.10 Summary

In this chapter we have given a series of small examples where we have modelled selected parts of the world: concrete phenomena of our everyday life, such as switches and light bulbs, traffic lights, agents with imperfect information about cards, etc. Of course, our scenarios are toy examples: they are simplistic and 'real life' is far more complex, involving much more information. However, our examples have the advantage of being at the same time complex enough to get a grasp of some important basic concepts and simple enough to understand what is going on. We could model more complex systems in the same way as we did in this chapter, such as an automatic control system of a network of subway trains, a nuclear power plant, computer programs, a particular body of law, etc.

All our examples have in common that they involve concepts that are fundamental in the modelling of many problems: time, events, actions, knowledge, belief, obligations, permissions, and objects and their relations. We have shown that theories of these concepts can be modelled by means of classes of graphs.

In the end of the present chapter we have explained how to build graphs with our tool LoTREC. When doing that we have met the LoTREC actions

[6] There is a subtle difference between a class and a set. All sets are classes, but not the other way round. For example, if we are looking for the set of all sets not containing themselves then we get into the troubles of the set theory paradox: this is an example of a class that is not a set.

createNewNode, link and add. Then we have given a formal definition of Kripke models that will serve as a base for the rest of this book.

Modelling a phenomenon with a graph is not merely a matter of graphical representation. As we shall see in the next chapter, we can also describe these graphs in a formal language and we can check whether a given graph satisfies some properties. We can also check whether all graphs of a given class satisfying some property A also satisfy some other property B. In other words, we can reason about properties in a formal language. The above concepts of time, events, actions, knowledge, and belief will play a particular role in that language: together with the boolean connectives of propositional logic they will have the status of *logical connectives*.

Chapter 2
Talking About Graphs

In the first chapter we have seen that graphs are a tool that allows us to model some important concepts. Up to now we have talked about such graphs in natural language. In this chapter we are going to introduce a particular *formal language* in order to do that: the language of modal logics.

For many reasons it is highly desirable to use a formal language instead of plain English: on the one hand, formal languages are more constrained than natural languages, thus more able to avoid ambiguities; on the other hand, a formal language allows one to rigorously verify each property that can be expressed in it. In particular, we may use a computer in order to verify these properties in a automatic or semi-automatic manner. To do so we have to communicate with the computer in a well-defined language.

All the examples of the preceding chapter could be presented in the language of *first-order logic*: all we need for that are unary predicates for the node labels and binary predicates for the edge labels. The former have a single node as an argument, and the latter have two nodes as arguments: a source and a target node. For example, the graph of Fig. 1.3 (Sect. 1.1, page 3) can be described in first-order logic by the formula

$$\mathsf{LightOff}(w) \land \mathsf{LightOn}(u) \land \mathsf{Toggle}(w, u) \land \mathsf{Toggle}(u, u)$$

where w and u are names for the left and the right node of the graph. Moreover, the sentence "When the light is off then toggling the switch turns the light on" describing that graph can be written in first-order logic as

$$\forall w\big(\mathsf{LightOff}(w) \to \forall u\big(\mathsf{Toggle}(w, u) \to \mathsf{LightOn}(u)\big)\big)$$

where $\forall$ is the universal quantifier "for all" and $\to$ is the material implication "implies" or "if... then..." This formula may be read: "for every node w, if w is labelled LightOff then for every node u, there is a link from w to u labelled Toggle implies that u is labelled LightOn."

O. Gasquet et al., *Kripke's Worlds*, Studies in Universal Logic,
DOI 10.1007/978-3-7643-8504-0_2, © Springer Basel AG 2014

Similarly, part of the information in Fig. 1.8 of Sect. 1.3 is formally expressed by

$$\forall w\big(\mathsf{Black}(w) \to \big(\forall u\big(R_{\mathsf{Bob}}(w,u) \to \mathsf{Black}(u)\big)\big) \\ \wedge \big(\forall u\big(R_{\mathsf{Ann}}(w,u) \to \big(\mathsf{Black}(u) \vee \mathsf{Red}(u)\big)\big)\big)\big)$$

However, in this book we shall not use the language of first-order logic. There are several reasons for that. First, if we express in the language of first-order logic statements about e.g. the agents' knowledge in the card game of Sect. 1.3 then we have to mention explicitly the nodes. This makes us write down formulas such as

$$\exists w'\big(R_{\mathsf{Ann}}(w,w') \wedge \mathsf{Red}(w')\big) \wedge \exists w'\big(R_{\mathsf{Ann}}(w,w') \wedge \neg\mathsf{Red}(w')\big)$$

in order to express that Ann does not know whether the card is red or not, and

$$\forall w'\big(\mathsf{Bob}(w,w') \to \big(\exists w''\big(\mathsf{Ann}(w',w'') \wedge \mathsf{Red}(w'')\big) \\ \wedge \exists w''\big(\mathsf{Ann}(w',w'') \wedge \neg\mathsf{Red}(w'')\big)\big)\big)$$

in order to express that Bob knows that. Such formulas are somewhat cumbersome. More importantly, the second reason for not choosing first-order logic is that automated reasoning in that logic is notoriously difficult: it is undecidable.[1]

We are going to describe graphs in another, simpler language. That language has neither variables nor quantifiers: instead, there are *modal connectives*. These connectives are also called *modal operators* or *modalities*. The language is therefore called a modal language. For example, we describe the two above statements about Ann's ignorance by means of the following expressions:

$$\neg\mathsf{K}_{\mathsf{Ann}}\mathsf{Red} \wedge \neg\mathsf{K}_{\mathsf{Ann}}\neg\mathsf{Red}$$

$$\mathsf{K}_{\mathsf{Bob}}(\neg\mathsf{K}_{\mathsf{Ann}}\mathsf{Red} \wedge \neg\mathsf{K}_{\mathsf{Ann}}\neg\mathsf{Red})$$

This is clearly more natural, more intuitive and more concise than the above first-order logic formulas. The modal operators of these formulas are $\mathsf{K}_{\mathsf{Bob}}$ and $\mathsf{K}_{\mathsf{Ann}}$. They are read "Bob knows" and "Ann knows." More abstractly, we write expressions such as $\mathsf{K}_I A$, read "I knows that A," and $\neg\mathsf{K}_I A$, read "I does not know that A," where I is an *agent* and A is a formula. The expressions $\mathsf{K}_I A$ and $\neg\mathsf{K}_I A$ are themselves also formulas.

The sentence "I does not know that A" means that $\neg A$, i.e., the falsehood of A, is actually possible for I. In other words, it means that I can imagine a state in which $\neg A$ holds. On the contrary, "I knows that A" means that A holds in all states that I can imagine. The states that I can imagine are all those states that I cannot distinguish from the real state by the information that I has. In a Kripke model, the accessibility relation $R(I)$ for the modality K_I—made up by all those edges that are

[1] More precisely, we here refer to the satisfiability problem of first-order logic. This is the famous Church-Turing Theorem [Kle67]. We will explain in the end of the chapter what satisfiability and decidability mean.

labelled I—models what I cannot distinguish: we have $\langle w, u\rangle \in R(I)$ if and only if I is unable to distinguish w from u.

The identification of "I knows A" with "A holds in all states that I can imagine" makes explicit that the modal language involves quantification over nodes, just as the language of first-order logic. The main difference is that graph nodes do not occur explicitly in the modal language. Moreover, quantification in modal logic is always *relativised*: it is not over all nodes, but only over those that are accessible from the current node via some edge. From the point of view of first-order logic, the modal language can be identified with a *fragment* of the language of first-order logic.[2] In that sense, the modal language is weaker than that of first-order logic.

The modal language can actually be located halfway between the language of propositional logic and the language of first-order logic: contrarily to the language of propositional logic and just as that of first-order logic it allows for *quantification*; and contrarily to the language of first-order logic it only allows for a *restricted* form of quantification in that modal operators are relativised quantifiers.

In many cases the expressivity provided by modalities is sufficient. As far as our examples are concerned, they allow us to express things such as "The light is on and after every possible performance of the toggling action the light is off," "There is a state that is possible for Ann where the card is red," "All next states after red are green," "After having opened the door, the door is opened and I know that the sun is shining," "If I draw a card then it is possibly red," "If I toss a coin then the result is necessarily heads or tails." We shall see more examples of modalities in the sequel.

We start this chapter by showing how to build complex formulas from boolean and modal connectives, leading to the definition of a general language for modalities (Sect. 2.1). Then we show how we can define modal languages in LoTREC (Sect. 2.2). After that, we formally state what it means that a formula is true at a node of a graph (Sect. 2.3) and show how we can check this in LoTREC (Sect. 2.4). Finally, we formally define several reasoning problems consisting in checking the truth of formulas in graphs and classes of graphs (Sect. 2.5) and briefly mention axiomatisations (Sect. 2.6) and the complexity of satisfiability (Sect. 2.7).

2.1 The Formal Language

We are now going to define a general formal language in which we can talk about concepts such as time, knowledge, belief, and action by means of modal operators. The elements of our language are called *formulas*. These formulas talk about what is true at a given node of a graph: they are going to be either true or false there.

---

[2]That fragment is part of a larger fragment of the language of first-order logic that is called theguarded fragment [AvBN98]. Remarkably and contrarily to the satisfiability problem for formulas of the entire language of first-order logic, the satisfiability problem for formulas of that fragment is decidable.

2.1.1 Atomic Formulas

First of all, we want to be able to describe basic facts. For instance, simple sentences such as "The light is on," "The card is red," and "The card is black" are directly represented by expressions such as LightOn, Red and Black. We consider each such expression as a single symbol and suppose that it cannot be divided into smaller parts: we say that it is an *atomic formula* (also called a propositional variable). You may think of atomic formulas as the smallest sentences you can write in our formal language.

We group atomic formulas in the *set of atomic formulas* $\mathcal{P}$. The elements of $\mathcal{P}$ are the basic node labels in our graphs.

2.1.2 Boolean Connectives

Atomic formulas can be combined by *boolean connectives* in several ways, resulting in complex formulas. The main boolean connectives are: falsum $\bot$, read "false;" negation $\neg$, read "not;" conjunction $\wedge$, read "and;" and disjunction $\vee$, read "or." These connectives allow us to write for example:

- $\neg$LightOn for "The light is not on" or "The light is off;"
- S1Up $\wedge$ LightOn for "Switch 1 is up and the light is on;"
- S1Up $\vee$ S2Up for "Switch 1 is up or switch 2 is up."

The formula $\neg$LightOn is a *literal*: a literal is either an atomic formula P from $\mathcal{P}$ or the negation $\neg P$ of some P from $\mathcal{P}$.

In the above examples, boolean connectives combine atomic formulas. But we can also put together non-atomic formulas:

- (S1Up $\wedge$ S2Up) $\wedge$ LightOn for "Both switches are up and the light is on;"
- (S1Up $\wedge$ S2Up) $\wedge$ $\neg$LightOff for "Both switches are up and the light is not off;"
- $\neg$(LightOn $\wedge$ LightOff) for "It is not the case that the light is both on and off;"
- ($\neg$LightOn) $\wedge$ $\neg$LightOff for "The light is neither on nor off;"
- (LightOn $\vee$ LightOff) $\wedge$ $\neg$(LightOn $\wedge$ LightOff) for "The light is either on or off."

Observe that we have used parentheses in order to be able to read formulas in an unambiguous way. We will say more about this in Sect. 2.1.7.

Other Boolean Connectives According to the semantics of propositional logic, $A \vee B$ has the same meaning as $\neg((\neg A) \wedge (\neg B))$. We could therefore restrict the set of boolean connectives to $\neg$ and $\wedge$.

Instead of fewer boolean connectives we may as well use more, such as $\top$ ("True"), $\rightarrow$ ("If... then...", or "Implies"), $\leftrightarrow$ ("Is equivalent to"), and $\oplus$ ("Either... or..."). However, they can be defined in terms of $\bot$, $\neg$, $\wedge$, and $\vee$:

- $\top$ is defined as $\neg\bot$;

- $A \to B$ is defined as $(\neg A) \vee B$;
- $A \leftrightarrow B$ is defined as $(A \to B) \wedge (B \to A)$;
- $A \oplus B$ is defined as $(A \vee B) \wedge \neg(A \wedge B)$.

We will use these abbreviations throughout the book.

One may even think of connectives involving more than two formulas, such as the ternary "Either..., or..., or..." However, all of them can be defined in terms of the above ones. This is the case because the set of boolean connectives is *complete*: any boolean function can be captured by a combination of $\bot$, $\neg$, $\wedge$, and $\vee$.

2.1.3 Modal Connectives

Atomic propositions and boolean connectives allow us to talk about the labels of *a single* graph node. But we also want to talk about several nodes. For instance, we may want to say that the light is off *in the current state* and on *in the state that obtains after toggling the switch*; or we may want to say that the card is red *in the actual state*, and that *there is a state that is possible for Bob where the card is not red*; etc. In order to be able to do that we introduce *modal connectives*.

The basic form of a modal connective is either $[I]$ or $\langle I \rangle$, where I is an *edge label* from some set of edge labels $\mathcal{I}$ that we suppose to be given.

When $\mathcal{I}$ is a singleton $\{I\}$, i.e., when there is just one edge label I, then we in general write $\Box$ and $\Diamond$ instead of $[I]$ and $\langle I \rangle$. In older texts, instead of $\Box$ and $\Diamond$ the reader may encounter the notation L and M. This is in particular the case when the terms of necessity and possibility are understood in the strict sense (as opposed to the large sense, encompassing in particular epistemic and deontic necessity and possibility). In this case we also talk about *alethic operators*.

The modal connective $[I]$ together with a formula A makes up a new formula $[I]A$. We read that formula "For all I-successors (of the current node), A is true" or "A is necessary w.r.t. I". We therefore call $[I]$ a *necessity operator*.

The modal connective $\langle I \rangle$ together with A makes up a new formula $\langle I \rangle A$, read "For some I-successor (of the current node), A is true" or "A is possible w.r.t. I". We therefore call $\langle I \rangle$ a *possibility operator*.

The readings we have given are generic and may be altered in some contexts. Here are some examples:

- When reasoning about knowledge we read for example [Ann]Red as "Ann knows that the card is red" and we read ⟨Ann⟩Red as "It is compatible with Ann's knowledge that the card is red;"
- When reasoning about programs and actions we read for example [Toss]Heads as "After every execution of the tossing action the coin is heads" and we read ⟨Toss⟩Heads as "After some execution of the tossing action the coin is heads."

Actually the notation $[I]$ and $\langle I \rangle$ for modal operators is also generic, and we sometimes choose more mnemonic symbols in order to improve the readability of

Table 2.1 Typical modal connectives and their reading

Formula	Typical reading	Modality
[Toss]Heads	"After *every* possible execution of the action of tossing a coin, heads will lie face up"	dynamic
⟨Toss⟩Heads	"After *some* possible execution of the action of tossing a coin, heads will lie face up"	dynamic
$\mathsf{K}_{\mathsf{Ann}}$Red	"Ann knows that the card is red"	epistemic
$\hat{\mathsf{K}}_{\mathsf{Ann}}$Red	"It is possible for Ann that the card is red"	epistemic
$\mathsf{B}_{\mathsf{Ann}}$Red	"Ann believes that the card is red"	doxastic
O_I Stop	"It is obligatory that I stops"	deontic
P_I Stop	"It is permitted that I stops"	deontic
X Red	"The traffic light will be red next"	temporal
G Red	"The traffic light will be red henceforth"	temporal
F Red	"The traffic light will eventually be red"	temporal
Red U Green	"The traffic light will be red until it turns green"	temporal

formulas with several modalities. This is particularly useful when there are modalities of different kinds, such as several agent names and several program names. For example, consider a language whose set of edge labels $\mathcal{I}$ is made up of a temporal 'next' label X plus a set of edge labels $\mathcal{I}_1$ denoting agents, with $\mathsf{X} \notin \mathcal{I}_1$. Then we typically write $\mathsf{X}A$ instead of $[\mathsf{X}]A$ in order to highlight that the nature of the temporal operator is different from that of the epistemic operators $[I]$ with $I \in \mathcal{I}_1$. Likewise, consider the case of edge labels that are of two different kinds: agents and actions. Let $\mathcal{I} = \mathcal{I}_1 \cup \mathcal{I}_2$ where $\mathcal{I}_1$ is the set of agents and $\mathcal{I}_2$ is the set of actions and where $\mathcal{I}_1$ and $\mathcal{I}_2$ are disjoint, i.e., $\mathcal{I}_1 \cap \mathcal{I}_2 = \emptyset$. We distinguish the two kinds of modal operators by varying notation: we write the epistemic operators K_I, one per agent $I \in \mathcal{I}_1$, and we write the dynamic operators $[a]$, one per action $a \in \mathcal{I}_2$. Even when the knowledge operator is the only modal operator then $[I]$ and $\langle I \rangle$ are typically written K_I and $\hat{\mathsf{K}}_I$, i.e., a 'K' with a *hat*, that may be pronounced "k-hat." Some examples of customary writings of modal operators are collected in Table 2.1.

Remark 3 Note that we did not impose any restriction on the set of edge labels $\mathcal{I}$. It may in particular be made up of formulas again. The temporal operators 'until' and 'since' are examples of operators that are indexed by formulas. The formula $[\mathsf{U}A]B$ is read "A until B," and the formula $[\mathsf{S}A]B$ is read "A since B." Just as other so-called binary modal operators, such formulas are usually written $A \mathrel{\mathsf{U}} B$ and $A \mathrel{\mathsf{S}} B$ instead of $[\mathsf{U}A]B$ and $[\mathsf{S}A]B$, as done in the last line of Table 2.1.

Together, boolean and modal connectives allow us to express a lot of things. For example, $\langle\mathsf{Open}\rangle\top$ can be read "The action of opening the door is possible," and $[\mathsf{Open}]\bot$ can be read "The door cannot be opened." Here are some more examples. The sentence "The light is off and after toggling the switch it will be on" can be expressed by the formula

$$\neg\mathsf{LightOn} \wedge [\mathsf{Toggle}]\mathsf{LightOn}$$

The sentence “Ann knows whether the card is red or not” can be written

$$\mathsf{K}_{\mathsf{Ann}}\mathsf{Red} \vee \mathsf{K}_{\mathsf{Ann}}\neg\mathsf{Red}$$

The sentence “Ann does not know whether the card is red, but knows that Bob does” gets

$$\neg(\mathsf{K}_{\mathsf{Ann}}\mathsf{Red} \vee \mathsf{K}_{\mathsf{Ann}}\neg\mathsf{Red}) \wedge \mathsf{K}_{\mathsf{Ann}}(\mathsf{K}_{\mathsf{Bob}}\mathsf{Red} \vee \mathsf{K}_{\mathsf{Bob}}\neg\mathsf{Red})$$

The sentence “Toggling the switch twice does not change the state of the light” gets[3]

$$\big(\mathsf{Light} \to [\mathsf{Toggle}][\mathsf{Toggle}]\mathsf{Light}\big) \wedge \big(\neg\mathsf{Light} \to [\mathsf{Toggle}][\mathsf{Toggle}]\neg\mathsf{Light}\big)$$

Finally, “Next the light will be either red or green” gets[4]

$$\mathsf{X}\big((\mathsf{Red} \vee \mathsf{Green}) \wedge \neg(\mathsf{Red} \wedge \mathsf{Green})\big)$$

Not any sequence of atomic formulas and connectives makes up a formula: the construction has to follow some rules that we are going to detail in the sequel.

2.1.4 Duality of Modal Connectives

We have seen that the modal connectives come in two kinds: necessity operators $[I]$ and possibility operators $\langle I\rangle$. Both quantify over all nodes that are accessible via I edges: the former is a ‘for all’ quantifier (just as the universal quantifier $\forall$ of first-order logic), while the latter is a ‘there is’ (just as the existential quantifier $\exists$).

We have noted in Sect. 2.1.2 that one may consider a language whose only boolean connectives are $\neg$ and $\wedge$. As to the modal connectives, it will turn out that $\langle I\rangle A$ has the same meaning as $\neg[I]\neg A$. We may therefore consider the former as an abbreviation of the latter, just as we might consider $A \vee B$ to be an abbreviation of $\neg(\neg A \wedge \neg B)$. (We are actually going to do so in Sect. 4.7 in order to shorten the presentation.) The other way round, we could as well consider $[I]A$ to be an abbreviation of $\neg\langle I\rangle\neg A$. The connectives $[I]$ and $\langle I\rangle$ are called *dual*.[5] This is the same duality as in first-order logic, where $\forall x A$ has the same meaning as $\neg\exists x\neg A$ and $\exists x A$ has the same meaning as $\neg\forall x\neg A$.

[3] Remember that our primitive boolean connectives are $\bot$, $\neg$, $\wedge$, and $\vee$ and that the formula $A \to B$ abbreviates $\neg A \vee B$.

[4] Using abbreviations this could also be written as $\mathsf{X}(\mathsf{Red} \leftrightarrow \neg\mathsf{Green})$ or $\mathsf{X}(\mathsf{Red} \oplus \mathsf{Green})$.

[5] Note that duality is no longer the case if we replace the boolean connectives for intuitionistic connectives (cf. see Sect. 5.3): just as $\forall x$ and $\exists x$ are no longer dual in intuitionistic predicate logic, $[I]$ and $\langle I\rangle$ are no longer dual in intuitionistic modal logics: they have different meanings there.

2.1.5 The Definition of Formulas

We now give our official definition of formulas that will remain the same throughout the book.

We suppose given a countable set of node labels $\mathcal{P}$ and a countable set of edge labels $\mathcal{I}$. In logic the former are also called *atomic formulas* (or propositional variables), while the latter are sometimes called modalities; we here call them indexes because they can be viewed as indexes of the modal operators $\Box$ and $\Diamond$. Note that we do not suppose that $\mathcal{P}$ and $\mathcal{I}$ are disjoint: they may have symbols in common.

The set of formulas is then built from these sets by means of the logical operators.

Definition 4 (Formula) Let $\mathcal{P}$ and $\mathcal{I}$ be countable sets. The set of formulas $\mathcal{F}or$ is the smallest set such that:

1. $\mathcal{P} \subseteq \mathcal{F}or$;
2. $\bot \in \mathcal{F}or$;
3. if $A \in \mathcal{F}or$ then $\neg A \in \mathcal{F}or$;
4. if $A, B \in \mathcal{F}or$ then $(A \wedge B) \in \mathcal{F}or$;
5. if $A, B \in \mathcal{F}or$ then $(A \vee B) \in \mathcal{F}or$;
6. if $A \in \mathcal{F}or$ and $I \in \mathcal{I}$ then $[I]A \in \mathcal{F}or$;
7. if $A \in \mathcal{F}or$ and $I \in \mathcal{I}$ then $\langle I \rangle A \in \mathcal{F}or$.

The first clause $\mathcal{P} \subseteq \mathcal{F}or$ means that each atomic formula is a formula. For instance, Black, LightOn and Red are formulas. The second clause says that if we prefix a formula by means of the negation connective $\neg$ then we obtain a formula. The third and fourth clauses say that if we connect a formula with another formula by means of the connectives $\wedge$ or $\vee$ then we obtain a formula. For instance, since LightOn and Red are formulas, (LightOn $\wedge$ Red) is a formula too. The last two clauses say that if A is a formula and I is an index then $[I]A$ and $\langle I \rangle A$ are formulas. For instance (LightOn $\wedge$ Red) being a formula and Open being an index, $[Open]$(LightOn $\wedge$ Red) is a formula.

When $\mathcal{I}$ is empty then our language is simply that of propositional logic. When $\mathcal{I}$ is a singleton then we call the language *monomodal*. If $\mathcal{I}$ contains more than one element then we call the language *multimodal*.

Remark 4 As we have mentioned in Sects. 2.1.2 and 2.1.4, we might either consider more connectives, or fewer. Actually LoTREC is very flexible and allows for languages that are built from *any* set of connectives. We might for instance define a boolean language whose only connectives are $\top$ ('true') and $\oplus$ ('exclusive or').

Remark 5 As we have said in Remark 3, there are modal connectives having two formulas as arguments, such as the temporal operators U ('until') and S ('since'). In order to fit in the above definition, we have to replace the usual notation AUB for the 'until' operator by $[\mathsf{U}A]B$ (but we note already that LoTREC is more flexible than that and allows us to display this in the standard way). These operators fit in the above schema because we imposed no constraints on the set of edge labels $\mathcal{I}$; it may in particular contain formulas. However, when the indexes of $\mathcal{I}$ are not all atomic but have a structure then their rigorous definition requires an inductive construction by means of index connectives. If moreover the indexes may contain formulas (as is the case of the U operator of **LTL**) then one needs a definition of formulas and indexes by mutual induction. We do not go into the details of such definitions here and rather refer the reader to the formal definition of the language of **LTL** in Sect. 7.1.

Another example of a complex index structure is provided by the programs of propositional dynamic logic **PDL**, where complex programs may in particular be sequential and nondeterministic composition of programs as well as tests of the truth of a formula: the program 'A?' succeeds if A is true and fails otherwise. Just as for **LTL**, the rigorous definition of the language requires an inductive construction (see Sect. 7.2).

The Connectives of Description Logics The connectives of description logics are written in a way that differs from what we have seen up to now. While the negation symbol takes the same form, conjunction and disjunction are not written $\wedge$ and $\vee$ but $\sqcap$ and $\sqcup$, and modal operators are not written $[I]$ and $\langle I \rangle$ but $\forall I$ and $\exists I$. Moreover, there is a convention to present the logics using A for atomic formulas, C for formulas ("concepts"), and R for indexes ("relations"). The formula ("concept") $\forall R.C$ reads "every object that is related to the actual object (the object currently under consideration) by R has property C," and $\exists R.C$ reads "there is an object that is related to the actual object by R and that has property C." For example,

$$\mathsf{female} \sqcap \forall \mathsf{ParentOf}.\mathsf{Male}$$

says that the object under consideration is female and that all her children are male.

Description logics also have some more constructions allowing one to state that a node has some (possibly complex) label and that some concept is included in another concept. Some extensions also allow for world names in the language. We do not state them here and restrict our attention to those mirroring the modal language. We just remark that names for worlds are also encountered in hybrid logics [AtC06] which we introduce in Sect. 4.8.

2.1.6 Analysing a Formula: Subformulas, Formula Length, Arity

The formulas from the set $\mathcal{P}$ are called *atomic* because they cannot be decomposed into smaller formulas. In contrast, $(\mathsf{Red} \wedge \mathsf{LightOn})$ is not atomic but complex and can be decomposed into Red and $\mathsf{LightOn}$. We say that Red and $\mathsf{LightOn}$ are *subformulas* of $(\mathsf{Red} \wedge \mathsf{LightOn})$. By convention, $(\mathsf{Red} \wedge \mathsf{LightOn})$ is also a subformula of itself. The formula $\mathsf{K}_{\mathsf{Ann}}(\mathsf{Red} \vee \mathsf{Black})$ consists of the connective $\mathsf{K}_{\mathsf{Ann}}$ governing the subformula $(\mathsf{Red} \vee \mathsf{Black})$; one also says that the latter is in the scope of the modal connective $\mathsf{K}_{\mathsf{Ann}}$. The subformula $(\mathsf{Red} \vee \mathsf{Black})$ has two other subformulas: Red

and Black, combined by the connective $\vee$. All of these formulas are subformulas of $\mathsf{K}_{\mathsf{Ann}}(\mathsf{Red} \vee \mathsf{Black})$.

To compute all the subformulas of a given formula we need an inductive definition. Here it is.

Definition 5 (Subformulas) The set of subformulas of a given formula A—denoted by $\mathrm{SF}(A)$—is inductively defined as:

$$\mathrm{SF}(P) = \{P\}, \quad \text{for } P \in \mathcal{P}$$
$$\mathrm{SF}(\bot) = \{\bot\}$$
$$\mathrm{SF}(\neg B) = \{\neg B\} \cup \mathrm{SF}(B)$$
$$\mathrm{SF}(B \wedge C) = \{B \wedge C\} \cup \mathrm{SF}(B) \cup \mathrm{SF}(C)$$
$$\mathrm{SF}(B \vee C) = \{B \vee C\} \cup \mathrm{SF}(B) \cup \mathrm{SF}(C)$$
$$\mathrm{SF}\big([I]B\big) = \big\{[I]B\big\} \cup \mathrm{SF}(B)$$
$$\mathrm{SF}\big(\langle I\rangle B\big) = \big\{\langle I\rangle B\big\} \cup \mathrm{SF}(B)$$

For example:

$$\begin{aligned}\mathrm{SF}\big(\langle I\rangle\neg P\big) &= \big\{\langle I\rangle\neg P\big\} \cup \mathrm{SF}(\neg P)\\ &= \big\{\langle I\rangle\neg P\big\} \cup \{\neg P\} \cup \mathrm{SF}(P)\\ &= \big\{\langle I\rangle\neg P, \neg P, P\big\}\end{aligned}$$

Remark 6 Connectives that are defined as abbreviations such as material implication $\rightarrow$ have to be decomposed according to their definition. For example:

$$\begin{aligned}\mathrm{SF}(P \rightarrow Q) &= \mathrm{SF}(\neg P \vee Q)\\ &= \{\neg P \vee Q, \neg P, Q, P\}\end{aligned}$$

Remark 7 The definition of SF can be extended in the obvious way if we want to adopt other primitive connectives. For example, if we add material implication $\rightarrow$ as a primitive then

$$\mathrm{SF}(B \rightarrow C) = \{B \rightarrow C\} \cup \mathrm{SF}(B) \cup \mathrm{SF}(C)$$

and consequently $\mathrm{SF}(P \rightarrow Q)$ then becomes the set $\{P \rightarrow Q, P, Q\}$.

Remark 8 In the case of richer modal languages where the argument I of the modal operator $[I]$ is a formula or contains a formula the above definition has to be modified in order to also extract the subformulas contained in I.

Definition 6 (Length of a formula) Given a formula A, the length of A, denoted by $\text{length}(A)$, is inductively defined as:

$$\text{length}(P) = 1, \quad \text{for } P \in \mathcal{P}$$
$$\text{length}(\bot) = 1$$
$$\text{length}(\neg B) = 1 + \text{length}(B)$$
$$\text{length}(B \wedge C) = 1 + \text{length}(B) + \text{length}(C)$$
$$\text{length}(B \vee C) = 1 + \text{length}(B) + \text{length}(C)$$
$$\text{length}\big([I]B\big) = 1 + \text{length}(B)$$
$$\text{length}\big(\langle I\rangle B\big) = 1 + \text{length}(B)$$

Again, the definition may have to be by mutual induction if the set of edge labels has a richer structure.

The number of subformulas of a given formula A is bounded by the length of A: we have $\text{card}(\text{SF}(A)) \leq \text{length}(A)$. (This is proved by induction on the structure of A.)

Arity of Connectives The number of expressions that are combined by a given connective is called its *arity*. For example, the arity of $\wedge$ is 2 and the arity of $\neg$ is 1. As $[I]A$ is built from two arguments (I and A) the arity of [.] is 2. A connective of arity 1 is called *unary* and a connective of arity 2 is called *binary*.[6] We give the arity of some standard connectives in Table 2.2 below.[7]

2.1.7 Parenthesis Conventions

The formula $\mathsf{K}_{\mathsf{Ann}}(\mathsf{Red} \to \mathsf{Red})$ can be read just in one way: "Ann *knows* that if the card is red then it is red." Parentheses are important: suppose we omit them and just write $\mathsf{K}_{\mathsf{Ann}}\mathsf{Red} \to \mathsf{Red}$. This expression can be parenthesised in two different ways: in the way we gave above and as $(\mathsf{K}_{\mathsf{Ann}}(\mathsf{Red}) \to \mathsf{Red})$, read "If Ann knows that the card is red then it is red." The point is that according to the first reading $\mathsf{K}_{\mathsf{Ann}}$ is the main connective, whereas according to the second reading $\to$ is the main connective.

---

[6] The notion of arity in logic is the same as in arithmetic, where each operator has a fixed number of arguments; for example the arity of + is two, and the arity of $\sqrt{\ }$ is 1 (if it is the quadratic root; else it is 2).

[7] We note that in modal logic it is usual to only count the argument of [.] if [.] contains a formula, as is the case in linear-time temporal logic **LTL**. When the set of edge labels $\mathcal{A}ct$ is a singleton I then it is usually considered that [.] is a 0-ary operator. When the set of edge labels $\mathcal{A}ct$ is made up of atomic labels then it is usual to say that [.] is unary.

However, some of the parentheses can be omitted if we adopt the following conventions.

1. The outer parentheses are dropped. For example $(A \wedge (B \vee C))$ is written $A \wedge (B \vee C)$.
2. Since in propositional logic $A \wedge (B \wedge C)$ has the same meaning as $(A \wedge B) \wedge C$, we may drop the parentheses and just write $A \wedge B \wedge C$. This generalizes to an arbitrary number of conjuncts; e.g. $A_1 \wedge ((A_2 \wedge (A_3 \wedge A_4)) \wedge A_5)$ is written $A_1 \wedge A_2 \wedge A_3 \wedge A_4 \wedge A_5$. Similarly, we drop all parentheses inside a sequence of $\vee$ and write e.g. $A \vee B \vee C$ instead of $(A \vee B) \vee C$.
3. We may specify an order of strength between the connectives. First, it is usually considered that unary connectives bind stronger than binary connectives. For example $\neg A \wedge A$ is identified with $(\neg A) \wedge A$ and $\langle I \rangle A \wedge A$ is identified with $(\langle I \rangle A) \wedge A$. Second, it is commonly considered that $\wedge$ binds stronger than $\vee$: for example, $A \wedge B \vee C$ is identified with $(A \wedge B) \vee C$. Third, concerning the boolean connectives that are abbreviations it is commonly considered that $\vee$ binds stronger than $\rightarrow$ and that $\rightarrow$ binds stronger than $\leftrightarrow$.

These conventions do not allow us to entirely drop parentheses; for example it fails to disambiguate $A_1 \rightarrow A_2 \rightarrow A_3$. We might go further and stipulate that such sequences of $\rightarrow$ associate to the right; then $A_1 \rightarrow A_2 \rightarrow A_3$ is identified with $A_1 \rightarrow (A_2 \rightarrow A_3)$. However, a general problem with such conventions is that when formulas get bigger then it requires more and more cognitive efforts to identify the main connective.

2.2 Syntax Declaration in LoTREC

In the preceding section we have learned how to define a formal language on paper, starting from a basic set of node labels $\mathcal{P}$, a basic set of edge labels $\mathcal{I}$, and logical connectives. We now show how to define such a language in LoTREC, so that we can use the formulas as graph labels and indexes as edge labels.

We have already announced in Sect. 2.1.5 that the connectives $\neg$, $\wedge$, $\vee$, $[.]$, and $\langle . \rangle$ are a particular choice: LoTREC is more general and allows for *any* set of connectives. Moreover, LoTREC allows us to write the modal operators not just in the generic forms $[I]$ and $\langle I \rangle$, but as any expression, as soon as it has been declared as such in the definition of the language.

We have also said in Sect. 2.1.5 that the set of edge labels might have some structure and that edge labels may as well be formulas. In order to provide the most general framework, LoTREC adopts an even more general stance and does not distinguish between node labels and edge labels: the general principle is that the set of complex labels `L` is built by means of connectives from a set of *set of atomic* LoTREC *labels* $\texttt{L}_0$. We use *expression* as a general term for a—possibly complex—label.

In Sect. 1.8 we have introduced the typewriter font convention: whenever you encounter typewriter font then you know that we are talking about the language of

LoTREC, i.e., things that are either implemented or implementable in LoTREC. This extends to atomic and complex labels. It will also apply to things such as tableau rules and strategies that we will introduce later on.

2.2.1 Prefix Notation

In our official definition of formulas we used parentheses systematically. There is however another solution: we may write down a formula in prefix form, also called Polish notation. It consists in writing it down starting by the main connective of the formula. Then we successively write down each subformula, also in prefix form. For example, the formula $(\mathsf{K}_{\mathsf{Ann}}\mathsf{Red}) \land \mathsf{Red}$ is written in prefix form as $\land\ \mathsf{K}_{\mathsf{Ann}}\ \mathsf{Red}\ \mathsf{Red}$. In contrast, the formula $\mathsf{K}_{\mathsf{Ann}}(\mathsf{Red} \land \mathsf{Red})$ is written $\mathsf{K}_{\mathsf{Ann}} \land \mathsf{Red}\ \mathsf{Red}$.

Observe that there is no ambiguity in the prefix notation under the following conditions: connectives are identified as such, their arity is fixed, and there cannot be two connectives with the same name.

While prefix notation was popular in logic books in the past, nowadays presentations use infix notation and parentheses. In LoTREC we however chose to use prefix notation because this allows us to do without parsing mechanisms. For example, in order to enter the formula $P \land \neg Q$ the LoTREC user has to type `and P not Q`. This may burden the users at first, especially new users. However, we are confident that the LoTREC user will soon be familiar with it. Moreover, formulas written in prefix notation can be displayed in infix notation on the screen. This is explained in Sect. 2.2.5.

2.2.2 Defining Atomic Labels

By convention the elements of LoTREC's set of atomic labels L_0 are *words starting with a capital letter*. Examples are `Red`, `Black`, `LightOn`, `LightOff`, `Open`, `Toggle`, `Ann`, `Bob`, `I`, `P`, `Q`, `J`, etc.

2.2.3 Defining Connectives

The LoTREC user can define his own connectives at will. We are therefore not limited to the connectives $\bot$, $\neg$, $\land$, $\lor$, $[.]$, and $\langle.\rangle$. There is however one connective that is predefined: the 0-ary falsum connective $\bot$, written `False` in LoTREC and displayed `FALSE`. This is because graph nodes containing that connective have a particular status when we try to build a model: such graphs cannot be transformed into a legal model of the logic, and LoTREC signals this by colouring nodes that are labelled `FALSE`.

Table 2.2 Definitions of the standard connectives in LoTREC

Connective	Name	Arity	Displayed as	Priority	Associative
$\bot$	False	0	FALSE		
$\neg$	not	1	~_	5	yes
$\wedge$	and	2	(_ & _)	3	yes
$\vee$	or	2	(_ v _)	2	yes
[.]	neci	2	[_] _	4	yes
$\langle . \rangle$	posi	2	<_> _	4	yes

In order to define a connective the user has to enter the following information:

- Its *name*, which is any word *starting with a small letter*;
- Its *arity*, which is a natural number whose default value is 1;
- The way it is *displayed* on screen;
- Its *priority* in order to define the relative strength of binding, which is a natural number whose default value is 0;
- Whether parentheses can be dropped when the connective occurs twice in a row, the default being that they cannot. In the case of binary connectives this is the same as declaring whether it is *associative* or not.

In textbooks and articles, parentheses in iterated unary connectives are suppressed. For example, one writes $\neg\neg P$ and $[I][J]P$ instead of $\neg(\neg P)$ and $[I]([J]P)$. We stress that LoTREC does not do so by itself: the user has to declare explicitly that parentheses should not be displayed by ticking the 'associative' box in the LoTREC interface; otherwise, LoTREC will display them.

Table 2.2 contains the definitions of some standard connectives. We use the names not, and, and or for the boolean connectives and the names neci and posi for the generic, indexed modal connectives. (Priorities and associativity of the connective False may have given arbitrary values because it has no arguments.)

Beyond the two generic modal connectives [.] and $\langle . \rangle$, other connectives can be defined in LoTREC, such as the epistemic connective knows and the temporal connectives next, henceforth, eventually, until, and since. The temporal connectives next, henceforth, and eventually have arity 1, while until and since have arity 2. The epistemic connective knows typically has arity 2, taking as argument an agent and a formula; however, when only a single agent is considered then the agent argument is often dropped and knows typically has arity 1. Some examples are collected in Table 2.3. Table 2.4 lists the description logic connectives that we have introduced in Sect. 2.1.5.

The information about display, priority, and associativity is used by LoTREC in order to drop parentheses when displaying formulas. In Sect. 2.2.5 we explain how this works in detail.

Table 2.3 Definitions of supplementary connectives in LoTREC

Connective	Name	Arity	Displayed as	Priority	Associative
$\rightarrow$	imp	2	(_ -> _)	2	no
$\leftrightarrow$	eqv	2	(_ <-> _)	1	no
$\Box$	nec	1	[] _	4	yes
$\Diamond$	pos	1	<> _	4	yes
X	next	1	X _	4	yes
K	knows	2	K(_) _	4	yes
$\hat{K}$	knowshat	2	K^(_) _	4	yes
B	bel	2	B(_) _	4	yes
U	until	2	(_ U _)	4	no

Table 2.4 Definitions of description logic connectives in LoTREC

Connective	Name	Arity	Displayed as	Priority	Associative
$\bot$	empty	0	FALSE		
$\neg$	dnot	1	~_	4	yes
$\sqcap$	dand	2	(_ & _)	2	no
$\sqcup$	dor	2	(_ v _)	2	no
$\forall$	dforall	2	A _. _	4	yes
$\exists$	dexists	2	E _. _	4	yes

2.2.4 *Complex Labels in* **LoTREC**

The set of labels L of LoTREC's language is the smallest set such that:

- every element of L_0 is a label;
- if c is a connective of arity *n* and A1, ..., An are labels of L_0 then c A1 ... An is a label.

Some example formulas built from the connectives of Tables 2.2 and 2.3 are given in Table 2.5.

Remember that there is no ambiguity in the prefix notation in particular because connectives are identified by their initial small letter. We can therefore safely write c A1 ... An without parentheses, instead of c(A1,...,An).

The following LoTREC rule extracts the subformulas of any formula of the form $A \wedge B$ and adds them to the node containing it.

```
Rule BreakupConjunctions
 hasElement node and variable A variable B

 add node variable A
 add node variable B
```

Table 2.5 Infix notation vs. LoTREC's prefix notation: examples

Formula on paper	LoTREC syntax
$P \vee (Q \wedge R)$	`or P and Q R`
$(P \vee Q) \wedge R$	`and or P Q R`
$P \wedge \neg Q$	`and P not Q`
$\mathsf{K}_I(P \wedge \neg Q)$	`knows I and P not Q`
$\mathsf{K}_I P \wedge \neg Q$	`and knows I P not Q`
$\mathsf{K}_{\mathsf{Ann}}(\mathsf{Red} \vee \mathsf{Black})$	`knows Ann or Red Black`

It can be implemented in LoTREC in the same way as we did for the model building rule `BuildExampleGraph` in Sect. 1.7 of Chap. 1 (page 16): add the rule `BreakupConjunctions` in the rule tab and edit the default strategy in the strategy tab such that it calls `BreakupConjunctions`. Observe that in that rule `A` and `B` are variables for formulas because they are prefixed by the LoTREC keyword <u>`variable`</u>. In contrast, the atomic labels in the formulas to be decomposed are constants (starting with a capital letter).

Exercise 16 Using `hasElement` conditions and `add` actions, write LoTREC rules which decompose formulas of the form $\neg A$, $A \rightarrow B$, and $[I]A$ into subformulas, one per connective. Distinguish the case where the implication connective is primitive from the case where it is an abbreviation.

2.2.5 Displaying Formulas

LoTREC allows the user to specify how formulas are displayed on the screen. He may in particular choose to display formulas in the familiar infix notation.

The way a given n-ary connective is displayed can be defined by a string of arbitrary characters with exactly n underscores '_'.

Table 2.2 contains the standard display definitions for some connectives. Accordingly, the LoTREC formula `or Red Black` is displayed as `Red v Black` and the LoTREC formula `knows Ann Red` is displayed as `K(Ann) Red`.

Now we turn to the way LoTREC allows us to economise on some parentheses when displaying formulas.

As we have mentioned in Sect. 2.1.7, it is a common convention in logic that unary connectives bind stronger than conjunction, allowing one to write for example $\neg A \wedge B$ instead of $(\neg A) \wedge B$. This is declared to LoTREC by giving to the connective `not` a higher priority than to `and`. The priorities of some standard connectives are in Table 2.2. Note that no such information is required for zero-ary connectives such as $\bot$ (its LoTREC priority may be set to any value). Consider for example the formula $\mathsf{K}_i P \wedge Q \rightarrow R$, written in LoTREC's prefix notation as `imp and knows I P Q R`. When given the priorities of Table 2.2, LoTREC displays this as `K(I) P & Q -> R`. The fact that the negation `not` has priority higher

Table 2.6 Infix notation, LoTREC's prefix notation, and LoTREC's display: examples

Formula on paper	LoTREC syntax	LoTREC display
$P \vee (Q \wedge R)$	`or P and Q R`	`P v Q & R`
$(P \vee Q) \wedge R$	`and or P Q R`	`(P v Q) & R`
$\mathsf{K}_I(P \wedge Q)$	`knows I and P Q`	`K(I) (P and Q)`
$\mathsf{K}_I P \wedge Q$	`and knows I P Q`	`K(I) P and Q`
$\mathsf{K}_{\mathsf{Ann}}(\mathsf{Red} \vee \mathsf{Black})$	`knows Ann or Red Black`	`K(Ann) (Red or Black)`

than for example the modal operator `[]` makes that LoTREC displays the formula $\Box\neg P$ as `[] ~P` on the one hand, and displays the formula $\neg\Box P$ as `~([] P)` on the other.

Finally, if we stipulate that a connective is associative then parentheses of sequences of the same connective are dropped. For example, the standard conjunction and disjunction connectives are both associative, while the implication connective is not. In the LoTREC interface we have to tick the respective fields in the definition of these connectives. For example, under the declarations of Table 2.2, the formulas `and and P Q R` and `and P and Q R` are displayed in the same way, namely as `P & Q & R`. Unary connectives should be declared to be associative (even if this is a bit at odds with the usual sense of the term): we can drop parentheses from double negation $\neg(\neg A)$ because there is only one way of parenthesising $\neg\neg A$.

The way some of the example formulas of Table 2.5 are displayed is given in Table 2.6.

2.3 Truth Conditions

In Chap. 1 we modelled various situations with Kripke models. In this section we formally define how formulas are evaluated in a possible world of a Kripke model as defined in Sect. 1.6.

Remember that the set of connectives is $\neg$, $\wedge$, $\vee$, $[I]$, and $\langle I \rangle$.

Given a Kripke model M, a state w of M, and a formula A, we want to define what it means that A is true at w. For instance, suppose M models the functioning of a coffee machine and w represents the state "Your coin has been inserted." How can we evaluate whether the formula [selectCoffee]Coffee ("After having selected a coffee, I will have a coffee") is true?

Formally we write $M, w \Vdash A$ for "The formula A is true in the world w of the model M" and define the relation $\Vdash$ as follows.

Definition 7 (Truth conditions) The forcing relation $\Vdash$ between models, worlds, and formulas is inductively defined to be the smallest relation such that:

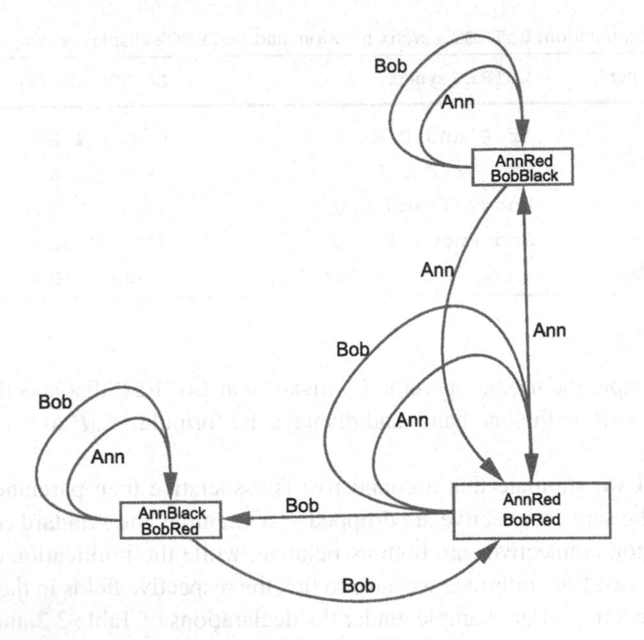

Fig. 2.1 Ann and Bob playing cards (possible world at top is w_0)

$M, w \Vdash P$ iff $w \in V(P)$, for $P \in \mathcal{P}$;

$M, w \nVdash \bot$;

$M, w \Vdash \neg A$ iff $M, w \nVdash A$;

$M, w \Vdash A \wedge B$ iff $M, w \Vdash A$ and $M, w \Vdash B$;

$M, w \Vdash A \vee B$ iff $M, w \Vdash A$ or $M, w \Vdash B$;

$M, w \Vdash [I]A$ iff for every world u, if $(w, u) \in R(I)$ then $M, u \Vdash A$;

$M, w \Vdash \langle I \rangle A$ iff there is a world u such that $(w, u) \in R(I)$ and $M, u \Vdash A$.

According to this definition, an atomic formula P is true in a world w of a model M iff the valuation of P contains w. A formula of the form $\neg A$ is true in w iff the formula A is false in w. A formula $A \wedge B$ is true in w iff both A and B are true in w, and $A \vee B$ is true in w iff A or B are true in w. Finally, the last two items say that formula $[I]A$ is true in w iff A is true in *all* the possible worlds u that are accessible from w by the relation $R(I)$; and formula $\langle I \rangle A$ is true in w iff A is true in *some* possible world u that is accessible from w by the relation $R(I)$.

Let us go back to Fig. 1.16 that describes a model M that we repeat as Fig. 2.1. Remember that the possible world at the top of the figure is called w_0. We can now state the following:

- $M, w_0 \Vdash \neg$BobRed;
- $M, w_0 \Vdash$ AnnRed $\wedge$ BobBlack;

- $M, w_0 \Vdash \neg(\mathsf{AnnRed} \wedge \mathsf{BobRed})$;
- $M, w_0 \Vdash [\mathsf{Ann}]\mathsf{AnnRed}$;
- $M, w_0 \Vdash \langle\mathsf{Ann}\rangle\neg\mathsf{BobRed}$; this is the case because there is one world accessible by $R(\mathsf{Ann})$ from w_0 (namely w_0 itself) where BobRed is false.

Observe that the boolean connectives $\neg$, $\wedge$, and $\vee$ are *truth-functional*: the truth value of a negation $\neg A$ is a function of the truth value of A, and the truth values of $A \wedge B$ and $A \vee B$ are functions of the truth value of their constituents A and B. In contrast, modal connectives are *not* truth-functional: the truth value of $[I]A$ and $\langle I\rangle A$ does not depend on the truth value of A.

Back to Duality According to our truth conditions, for every M and w we have

$$M, w \Vdash \langle I\rangle A \quad \text{iff } M, w \Vdash \neg[I]\neg A$$

$$M, w \Vdash [I]A \quad \text{iff } M, w \Vdash \neg\langle I\rangle\neg A$$

Therefore and as we have announced in Sect. 2.1.4, the notions of possibility and necessity are interdefinable: one may alternatively only use the necessity operator and consider the possibility operator to be an abbreviation, or the other way round. Often it is convenient to have both modal connectives as primitives of the language, which is what we do here.

2.4 How to Do Model Checking in LoTREC

We now explain how to use the procedure `Model-Checking-Multimodal` that comes with the LoTREC distribution. The procedure works for a language with modal operators `[I]` and `<I>`, written where `I` is some edge label. (If your set of indexes $\mathcal{I}$ is a singleton you might avoid typing the arguments of the modal operators and use the modal operators `[]` and `<>`; in that case you should apply the procedure `Model-Checking-Monomodal`.)

Let $M = (W, R, V)$ be a finite Kripke model and let w be a possible world from W. Let A be the formula of which we want to know whether $M, w \Vdash A$.

We open the ‘predefined logic’ `Model-Checking-Multimodal`[8] and build the model `M` by means of a LoTREC rule in the same way we did in Sect. 1.7. Moreover add the formula `isItTrue(A)?` to the root node `w` of `M`.

We illustrate model checking by an example. Let us check whether the formula $\langle R\rangle[R](P \vee Q)$ is true in the root `w` of the graph `M` depicted in Fig. 2.2, where `w` is the uppermost node of the graph.

[8] `Model-Checking-Multimodal` and `Model-Checking-Monomodal` are in LoTREC’s list of predefined logics, but they are not the tableaux procedure for a particular logic. (They are the only such exceptions in that list.)

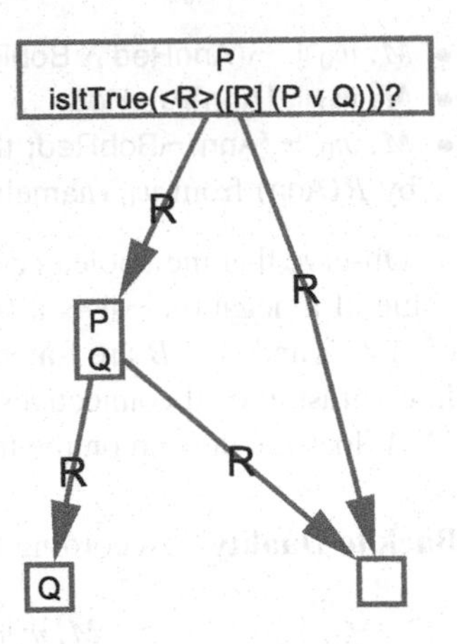

Fig. 2.2 An input of the model-checking problem

We build the LoTREC graph of the model M by means of the following rule:

```
Rule BuildExampleGraph
  createNewNode w
  createNewNode u
  createNewNode v
  createNewNode x
  link w u R
  link w v R
  link v u R
  link v x R
  add w P
  add v P
  add v Q
  add x Q
  add w isItTrue pos R nec R or P Q
```

where `isItTrue` is defined in LoTREC to be a unary connective that is displayed as `isItTrue(_)?`. This is not a modal operator of the logic, but an auxiliary connective that allows us to distinguish the formulas to be evaluated during the model checking process. Observe that the rule has finite length because M is finite: the set of nodes W to be created is finite and there is only a finite number of labels that have to be added to nodes and edges. Beyond the above rule `BuildExampleGraph` there are several other predefined rules in `Model-Checking-Multimodal`. Their job is to break up the formula to be evaluated in a first pass, and then to compute the truth value of each subformula in a second pass. How this is done is explained in detail in Chap. 6.

Once the model building rule has been entered you may push the "Build Premodels" button: LoTREC will then display the result depicted in Fig. 2.3. The result shows that $M, w \Vdash \langle R \rangle [R](P \vee Q)$, as indicated by the tag `[Yes]` that LoTREC has placed in `w` right after the input formula.

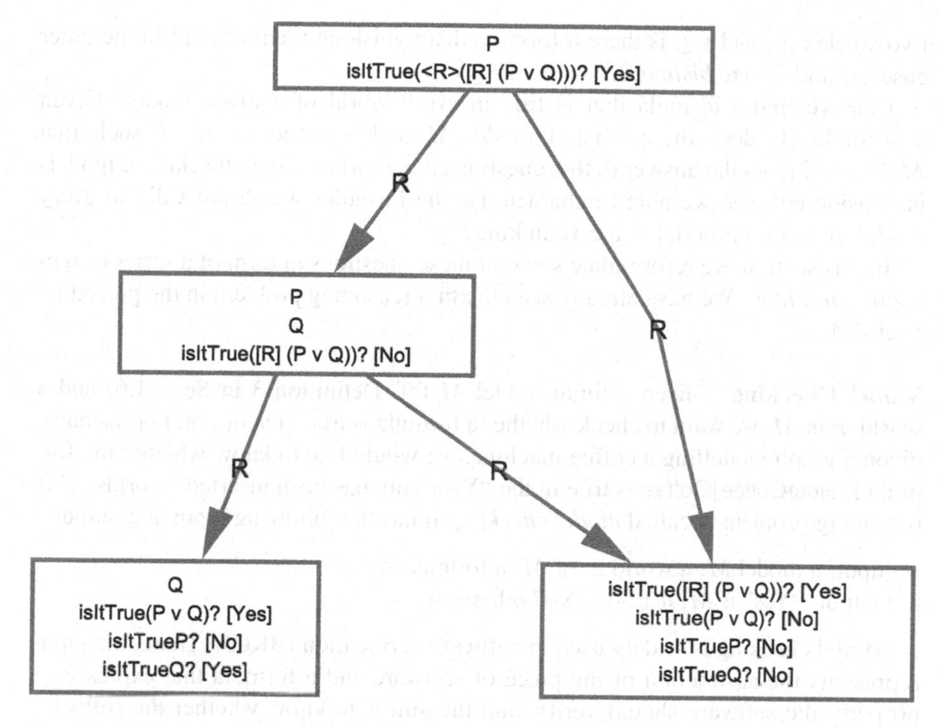

Fig. 2.3 The output of the model checking procedure

You may save the model: use "Save selected premodel" in the "Premodels" menu. You may reload the model and check another formula by changing the `BuildExampleGraph` rule appropriately.

Exercise 17 Let *M* be the graph of Fig. 1.4 in Sect. 1.1. Let w_0 be the upper possible world of that graph (where the light is on and both switches are down). Check whether

$$M, w_0 \Vdash [\mathsf{Toggle}_{\mathsf{S1}}]\mathsf{LightOn}$$

Now let w_1 be the right possible world of that graph (where the light is off, the first switch is up and the second switch is down). Check whether

$$M, w_1 \Vdash [\mathsf{Toggle}_{\mathsf{S1}}]\mathsf{LightOn}$$

2.5 Modal Logics and Reasoning Problems

Various questions may be asked about formulas and their truth values in worlds of Kripke models. For instance, does the truth value of a formula in a specific world of a model change when it is considered in another world of the same model? Given two worlds w_1 and w_2, is there a formula distinguishing them or not? In the latter case w_1 and w_2 are *bisimilar*.[9]

Can we find a formula that is true in every world of a given model? Given a formula A, does there exist a model M and a world w of M such that $M, w \Vdash A$? Does the answer to this question change when a specific class of models is considered? Are we able to characterise the formulas which are valid in every model, or in every model of a certain kind?

In this section, we reformulate some of these questions in form of a series of *reasoning problems*. We have already seen the first reasoning problem in the preceding Sect. 2.4.

Model Checking Given a finite model M (cf. Definition 3 in Sect. 1.6) and a world w in M, we want to check whether a formula is true in w or not. For instance, given a graph modelling a coffee machine, we would like to know whether the formula [selectCoffee]Coffee is true in the "Your coin has been inserted" worlds. This reasoning problem is called *model checking*. It has the following input and output.

- Input: a model M, a world w of M, a formula A;
- Output: "Yes" if $M, w \Vdash A$; "No" otherwise.

Model checking is widely used in software verification [BK08]. There, the input represents the abstraction of the piece of software and a formula that expresses a property the software should verify, and the aim is to know whether the software verifies that property or not.

Validity Checking Some formulas such as $P \vee \neg P$ are clearly true in every world of every Kripke model. Such formulas are called valid. In contrast, the atomic formula P is not valid: it is easy to come up with a Kripke model M and a world w of M such that P is false at w. Take e.g. a model M that is made up of a single possible world w, an empty accessibility relation and an empty valuation: clearly, $M, w \nVdash P$.

It is quite easy to see that $P \vee \neg P$, $[I]A \leftrightarrow \neg\langle I\rangle\neg P$, and $\langle I\rangle A \leftrightarrow \neg[I]\neg P$ are valid. Checking the validity of any formula is the following problem.

- Input: a formula A;
- Output: "Yes" if $M, w \Vdash A$ for every possible world w of every Kripke model M; "No" otherwise.

---

[9]More generally, given two models $M_1 = \langle W_1, R_1, V_1\rangle$ and $M_2 = \langle W_2, R_2, V_2\rangle$, a *bisimulation* between M_1 and M_2 is a relation $Z \subseteq W_1 \times W_2$ such that

- if $\langle w_1, w_2\rangle \in Z$ then for every atomic formula P, $w_1 \in V(P)$ iff $w_2 \in V_2(P)$;
- if $\langle w_1, w_2\rangle \in Z$ then for every edge label $I \in \mathcal{I}$,
  - if $\langle w_1, u_1\rangle \in R_1(I)$ then there is $u_2 \in W_2$ such that $\langle w_2, u_2\rangle \in R_2(I)$;
  - if $\langle w_2, u_2\rangle \in R_2(I)$ then there is $u_1 \in W_1$ such that $\langle w_1, u_1\rangle \in R_1(I)$.

Two pointed models M_1, w_1 and M_2, w_2 are bisimilar if there is a bisimulation Z between M_1 and M_2 such that $\langle w_1, w_2\rangle \in Z$.

This is the *validity problem in the class of all Kripke models*. It can also be posed w.r.t. a subset of the class of all Kripke models.

Logic of a Class of Frames, Logical Consequence We say that a formula A is valid in a class of frames $\mathbf{C}$ if $\langle W, R, V\rangle, w \Vdash A$ for every frame $\langle W, R\rangle \in \mathbf{C}$, every valuation V over $\langle W, R\rangle$ and every w in W.

The *logic of a class of frames* $\mathbf{C}$ is the set of formulas A such that A is valid in $\mathbf{C}$. To say that A is valid in $\mathbf{C}$ is therefore the same as saying that A belongs to the logic of $\mathbf{C}$.

Given two formulas A and B and a class of models $\mathbf{C}$, we say that B is a logical consequence of A in the class of models $\mathbf{C}$, noted $A \models_{\mathbf{C}} B$, if and only if for every model $M \in \mathbf{C}$ and every possible world w of M, if $M, w \Vdash A$ then $M, w \Vdash B$.[10]

Exercise 18 Prove that $A \models_{\mathbf{C}} B$ if and only if $A \to B$ is valid in $\mathbf{C}$.

Satisfiability Checking The formula P is clearly satisfiable: take e.g. a single possible world w, an empty accessibility relation R, and a valuation function V such that $V(P) = \{w\}$; then $\langle \{w\}, R, V\rangle \Vdash P$. In contrast, the formula $\neg P \wedge P$ is clearly *unsatisfiable*.

More generally, the *satisfiability checking problem* is that we want to check whether for a given formula there exists or not a Kripke model M and a world w in M such that the formula is true at w. Formally it has the following description.

- Input: a formula A;
- Output: "Yes" if there is a model M and a world w of M such that $M, w \Vdash A$; "No" otherwise.

One often only wants to check for existence of a model of a particular kind, satisfying some constraints on the accessibility relation or on the valuation function. These constraints restrict the search space to special classes of models: we are interested in validity in the class of models $\mathbf{C}$ under concern. We then talk about the *problem of satisfiability in* $\mathbf{C}$, where $\mathbf{C}$ is some special class of models. That class is part of the input of the satisfiability problem.

Suppose we have a procedure solving the satisfiability problem. If the answer for the input formula $\neg A$ is "No, there is no model M having a world w of M such that $M, w \Vdash \neg A$" then the validity problem for the formula A is also solved: the answer is: "Yes, A is valid." Conversely, when the answer is "Yes, there is a model M and a world w such that $M, w \Vdash \neg A$" then the answer for the validity problem is "No, A is not valid." In other worlds, a procedure for the satisfiability problem

[10] This is sometimes called the local consequence relation. The global consequence relation is defined as follows: B is a global logical consequence of A in the class of models $\mathbf{C}$ if and only if for every model $M \in \mathbf{C}$, if $M, w \Vdash A$ for every possible world w of M then $M, w \Vdash B$ for every possible world w of M. We do not consider global logical consequence in this book.

Both consequence relations can be generalised to possibly infinite sets of formulas S on the left side of the symbol of logical consequence $\models_{\mathbf{C}}$. The logic of $\mathbf{C}$ is *compact* if $S \models_{\mathbf{C}} B$ implies that there is a finite subset S' of S such that $S' \models_{\mathbf{C}} B$. All the logics of Chaps. 3, 4 and 5 are compact, while the logics **LTL** and **PDL** of Chap. 7 are not.

can be easily turned into a procedure for the validity problem. The converse is also the case. Interdefinability is moreover the case for the validity problem and for the satisfiability problem in a class of models.

Model Construction The output of the above tasks of model checking, validity checking, and satisfiability checking is just "Yes" or "No." However, one would often like to get an *explanation* of these answers: why "Yes" and why "No"? In other words, we would like to understand why a given formula is satisfiable or not. This typically happens when students learn a logic or when researchers investigate the properties of an existing logic or design a new logic. In these cases one is interested in the following problem.

- Input: a formula A;
- Output: a model M and a world w such that $M, w \Vdash A$ if A is satisfiable; '*unsatisfiable*' otherwise.

This is the *model construction* or *model finding problem.* A couple $\langle M, w \rangle$ such that $M, w \Vdash A$ explains why "Yes, A is satisfiable" is the case.

Suppose now that we are checking the validity of a formula A by running a satisfiability checking procedure on $\neg A$, as explained above. When the procedure successfully returns a couple $\langle M, w \rangle$ then this means that $M, w \Vdash \neg A$. According to the truth condition for the negation operator this means that $M, w \nVdash A$. We have therefore a *counter-model* for A which explains why A is invalid.

Just as for satisfiability checking and validity checking, model construction may be restricted to models from some particular class of models.

2.6 Modal Logics and Their Axiomatisations

As we have already said, the (modal) logic of a class of frames **C** is the set of formulas A such that A is valid in **C**. This set is in general infinite. The class **C** is typically infinite, too.[11] This contrasts with propositional logic where the number of rows of the truth table for a given formula A is bound by the length of A.

It is an old idea in logic to try to characterise the logic of a given class **C** in a finite way, viz. by means of a finite set of axioms and a finite set of inference rules. This can be traced back to Hilbert's program for the validities of arithmetic and even to Euclid's geometry. Once one has come up with such an axiomatisation one may prove its *theorems*: formulas that can be deduced from instances of the axiom schemas via the inference rules. That deduction is called a *proof* of that theorem. However, in order to justify such an axiomatisation two things have to be established in the first place: one has to check that every theorem is valid, and the other way round, one has to check that every valid formula is a theorem. The former is called *soundness* and the latter is called *completeness*.

[11] More precisely, the class may not be countable. In terms of set theory, that class may not be a set.

Table 2.7 Axiomatisation of modal logic **K**

CPL	an appropriate set of axioms for classical propositional logic
M	$\Box(A \wedge B) \rightarrow (\Box A \wedge \Box B)$
C	$(\Box A \wedge \Box B) \rightarrow \Box(A \wedge B)$
N	$\Box\neg\bot$
Dual	$\Diamond A \leftrightarrow \neg\Box\neg A$
MP	from A and $A \rightarrow B$ infer B
RE	from $A \leftrightarrow B$ infer $\Box A \leftrightarrow \Box B$

Table 2.7 contains a complete axiomatisation of the modal logic **K**: the logic of all Kripke frames. The name **K** is in honour of Saul Kripke. Axiom **M** is called the axiom schema of monotony; **C** is the axiom schema for conjunction; **N** is the axiom schema of necessity; **MP** is the rule of modus ponens; **RE** is the rule for equivalence. The term *axiom schema* (instead of axiom *tout court*) highlights that the symbols A and B occurring in them have to be understood as *schematic variables*: each of these variables may be uniformly replaced by any formula, where 'uniformly' means that it has to be the same formula for every occurrence of the same schematic variable.[12] The result is called an *instance* of that axiom.[13]

Let us put that axiomatics to work: we are going to prove that every instance of the so-called axiom **K**: $(\Box A \wedge \Box(A \rightarrow B)) \rightarrow \Box B$ is a theorem of modal logic **K**.

1. $(\Box A \wedge \Box(A \rightarrow B)) \rightarrow \Box(A \wedge (A \rightarrow B))$ (axiom **C**)
2. $(A \wedge (A \rightarrow B)) \leftrightarrow (A \wedge B)$ (**CPL**)
3. $\Box(A \wedge (A \rightarrow B)) \leftrightarrow \Box(A \wedge B)$ (from (2) by rule **RE**)
4. $\Box(A \wedge B) \rightarrow (\Box A \wedge \Box B)$ (axiom **M**)
5. $(\Box A \wedge \Box(A \rightarrow B)) \rightarrow \Box B$ (from (1), (3), (4) by a **CPL** rule)

When we say that the last line follows "from (1), (3), (4) by a **CPL** rule" we want to say that we apply a derived inference rule of classical propositional logic. That rule takes the form "from $X_1 \rightarrow X_2$, $X_2 \leftrightarrow X_3$, and $X_3 \rightarrow (X_4 \wedge X_5)$ infer $X_1 \rightarrow X_5$." It is applicable because formula (1) has the form $X_1 \rightarrow X_2$, formula (3) has the form $X_2 \leftrightarrow X_3$, and formula (4) has the form $X_3 \rightarrow (X_4 \wedge X_5)$. We note that in the light of our distinction between axiom and axiom schema (where the symbols A and B occurring in a schema are understood as schematic variables), the formula $(\Box A \wedge \Box(A \rightarrow B)) \rightarrow \Box B$ might perhaps better be called a theorem schema: any of its instances is also provable.

[12] We abuse our notation a bit here because we use the symbols A and B both in order to designate formulas and schematic variables. There should however never be a confusion because schematic variables only occur in axiom schemas, and such schemas will always be clearly identified as such.

[13] It is also possible to formulate axiomatisations in terms of axioms instead of axiom schemas. Such axioms have atomic formulas P and Q instead of schematic variables A and B. One then has to add the rule of uniform substitution which allows us to replace atomic formulas by other formulas.

Table 2.8 Some basic modal axioms

Axiom name	Axiom schema	Constraint
T	$\Box A \rightarrow A$	reflexivity
D	$\Box A \rightarrow \Diamond A$	seriality
Alt$_1$	$\Diamond A \rightarrow \Box A$	determinism
B	$A \rightarrow \Box \Diamond A$	symmetry
4	$\Box A \rightarrow \Box \Box A$	transitivity
5	$\Diamond A \rightarrow \Box \Diamond A$	euclideanity
.2 (or **G**)	$\Diamond \Box A \rightarrow \Box \Diamond A$	confluence
.3	$\Diamond A \wedge \Diamond B \rightarrow \Diamond(A \wedge B) \vee \Diamond(A \wedge \Diamond B) \vee \Diamond(B \wedge \Diamond A)$	right linearity

The logic of the class of reflexive frames is called **KT**. An axiomatisation of **KT** can be obtained from that of **K** by adding the axiom schema **T**: $\Box A \rightarrow A$ to it.

Let us show that $P \rightarrow \Diamond P$ is a theorem of modal logic **KT**.

1. $\Box \neg A \rightarrow \neg A$ (axiom **T**)
2. $A \rightarrow \neg \Box \neg A$ (from (1) by a **CPL** rule)
3. $\neg \Box \neg A \leftrightarrow \Diamond A$ (axiom **Dual**)
4. $A \rightarrow \Diamond A$ (from (2) and (3) by a **CPL** rule)

Observe that the first line is not exactly axiom **T**, but rather an instance of it.

Actually the schema $A \rightarrow \Diamond A$ may replace axiom **T** in the axiomatisation of **KT**.

Exercise 19 Prove that $\Box P \rightarrow \Diamond P$ is a theorem of **KT**.

We are now going to present several standard modal logics. We focus on classes of frames with a single accessibility relation, and therefore on a monomodal language, i.e., a language where the set of edge labels is a singleton. We have already seen that the logic of the class of all Kripke models is called **K** and that the logic of reflexive frames is called **KT**. In Table 2.8 below we list the axioms corresponding to the semantic constraints of Table 1.1 in Sect. 1.9.

The names of the modal logic of some subset of the constraints of Table 1.1 are well-established names in the literature. They are obtained by adjoining the respective list of axiom schemas to the letter **K**. For example, **KD** is the logic of serial frames, **KB** is the logic of symmetric frames, and so on. Sometimes **KD** is just called **D**, **KT** is just called **T** and **KB** is just called **B**. One may also add more than a single axiom to **K**: the axiom of reflexivity being **T** and the axiom of symmetry being **B**, the logic of the class of reflexive and symmetric Kripke models is called **KTB**.

Observe that every reflexivity relation is also serial; therefore the logics **KT** and **KDT** are the same. We also have **KT5** = **KDT5**, etc. We moreover have **KT45** = **KTB5**. There are some logics with historic names which differ from our nomenclature and which are often used: the traditional name of **KT4** is **S4**, that of **KT45** is **S5**, that of **KT4.2** is **S4.2**, that of **KT4.3** is **S4.3**.

Up to now we have only considered logics of a single relation. As to multiple relations, the logic of the class of all Kripke models having n edge labels is designated by $\mathbf{K}_n$. So the monomodal **K** is the same as $\mathbf{K}_1$, and $\mathbf{K}_2$ is the logic of two relations which do not satisfy any constraint. That class of these models is also written $\mathbf{K} \otimes \mathbf{K}$. Such a notation allows us to denominate other logics, such as that of the class of frames with two relations one of which is reflexive: it is written $\mathbf{K} \otimes \mathbf{KT}$. We may also have constraints involving two relations. A simple example is the inclusion of one relation in another. The logic of such frames is $\mathbf{K}_2 \oplus$ **Inclusion** $= (\mathbf{K} \otimes \mathbf{K}) \oplus$ **Inclusion**. Instead of introducing these classes here we will define them on the fly.

Our language being enumerable, we may use a finite axiomatisation in order to automatically enumerate all the theorems. However, this does not allow us to *decide* whether a given formula A is a theorem or not: if A is a theorem then the enumeration will reach it some day; but if it is not then the enumeration procedure will run forever. It is however possible to obtain a decision procedure for **C** validity when the logic of **C** moreover has the *finite model property*, abbreviated fmp. The latter says that for every formula A, if A is satisfiable in **C** then there is a finite $M \in \mathbf{C}$ and a world w in M such that $M, w \Vdash A$. Indeed, if the logic has the fmp then it basically suffices to enumerate in parallel the theorems of the axiomatisation on the one hand and the finite models of **C** on the other:[14] if A is a theorem then the former enumeration will reach it, and if A is invalid then the latter will reach a *countermodel*. (Remember that such decision procedure only works correctly if soundness and completeness have been proved beforehand.)

Note that one may as well try to directly come up with a proof of A by fiddling around with the axiom schemas and inference rules. Actually many of the proofs that can be found in textbooks are of that kind. They require quite some creativity, and this is an enterprise that is closer to craftsmanship than to an automated procedure. We have seen in some of the above proofs that one has to guess the 'right' instances of the axiom schemas to start with. This is typical for *Hilbert-style axiomatisations*. These are typically axiomatisations with many axioms and few rules: modus ponens plus one or two inference rules per modal operator. In contrast, the tableaux method—as well as sequent calculi and natural deduction calculi—can be viewed as having a single axiom and many rules.

Axiomatisations such as the above for logics **K** and **KT** provide a mechanical way to enumerate theorems and seems to provide a way for designing algorithms for automated reasoning. However, no algorithm for satisfiability checking or validity checking is based on a Hilbert axiomatisation. Indeed, even if the above decision procedure works in theory, it is however very inefficient. As we will see, tableaux provide for a more efficient procedure.

[14] Actually we should be able to enumerate the finite models of **C**. It is possible for all classes referred in this book. Nevertheless it is not always the case, for instance for this tricky example: the class of all models where the number of worlds n is such that the nth Turing machine does not halt.

2.7 A Note on Computational Complexity

The problem of satisfiability of formulas of first-order logic is undecidable. This contrasts with modal logics: as M. Vardi once formulated it, modal logic satisfiability problems are "surprisingly often decidable" [Var96].

In this section, we briefly recap the main results about the complexity of the satisfiability problem in modal logics.

The aim of computational complexity theory is to characterise the resources that are needed to solve a decision problem. A decision procedure solving the satisfiability problem needs two kinds of resources: time and memory.

The set of decision problems that can be solved by a deterministic algorithm working in time polynomial in the input is called P.

The set of decision problems that can be solved by a nondeterministic algorithm working in time polynomial in the input is called NP. Problems in the set NP can typically be solved by (nondeterministically) guessing a solution of polynomial size and then checking in polynomial time that the solution is correct. For instance, the satisfiability problem of classical propositional logic is in NP: for a given formula A, after guessing a truth-value for each atomic proposition occurring in A, we may check in polynomial time that the resulting truth-value assignment makes A true. It was proved by Stephen Cook that the satisfiability problem of classical propositional logic is among those problems of NP that are most difficult to solve: any other reasoning problem in NP can be transformed into a satisfiability problem of classical propositional logic, and so in polynomial time [Coo71]. For that reason we say that the satisfiability problem of classical propositional logic is NP-complete.

It has to be noted that while almost all researchers believe that the set of problems solvable in polynomial time is strictly included in NP, this has actually not been proven yet and remains a conjecture.

There are problems that cannot be solved in nondeterministic polynomial time and that are therefore beyond the set NP. Here are two of them:

- PSPACE is the set of decision problems that can be solved by an algorithm that only uses a polynomial amount of memory;
- EXPTIME is the set of decision problems that can be solved by an algorithm that only uses an exponential amount of time.

In the same way as for NP one can define PSPACE completeness and EXPTIME completeness. We have P $\subseteq$ NP $\subseteq$ PSPACE $\subseteq$ EXPTIME. We have already said that it is an open problem whether the inclusion P $\subseteq$ NP is strict; the same is the case for each of the three other inclusions.

Figure 2.4 classifies the most important modal logics according to the complexity of their satisfiability problem.

2.8 Summary

In this chapter we have finished setting the stage for this book: in order to talk about graphs we have defined a general formal language having parametrised modal

NP-complete	**S5, KD45, S4.3**
PSPACE-complete	**K, KD, KB, KT, K4, S4, LTL**
EXPTIME-complete	**PDL**

Fig. 2.4 Complexity of the satisfiability problem of different modal logics. (The logics **LTL** and **PDL** are defined in Chap. 7)

operators. We have then shown how to define a particular formal language on our tableaux platform LoTREC by listing the logical connectives together with their arity and the way they are displayed. We have then given a truth condition for each of these connectives: together, they allow one to decide whether a formula is true or false in a given world of a given model. We have finally presented four important reasoning problems related to the evaluation of formulas: model checking, validity checking, satisfiability checking, and model construction. However, we have not said how these problems can be solved. This is what we are going to do from the next chapter on.

Let us already note that for the problems of validity, satisfiability, and model building the search space is clearly very large: we look for an arbitrary model M and an arbitrary world in that model where the formula is true. It is *a priori* not clear how the size of such a model and the size of the search space of all relevant models could be bounded. This contrasts with e.g. the satisfiability problem for a formula A of propositional logic where it suffices to check the relevant propositional valuations, i.e., the set of subsets of the set of atomic formulas of A: there are only exponentially many such valuations, depending on the number of atomic formulas occurring in A, i.e., depending on the cardinality of the set $\mathcal{P} \cap \mathrm{SF}(A)$.

We have seen that a solution to the model construction problem also allows us to check satisfiability and validity, while the converse is not the case. We therefore focus in this book on a method for constructing models. We will start by a method searching for a model in the class of all Kripke models and then examine methods for various classes of models. These will be explained throughout the next chapters. We will get back to the model checking problem in Chap. 6.

Chapter 3
The Basics of the Model Construction Method

In the preceding chapter we introduced a formal language to talk about graphs: the language of modal logic. We have also said what the model checking task is: to find out whether a given formula is true at some node of a given graph. We have also stated another reasoning problem that is going to be central in this book: model construction. We recall that it takes the following form:

- Input: a formula A;
- Output: a model M and a world w such that $M, w \Vdash A$ if A is satisfiable; '*unsatisfiable*' otherwise.

Such a task is generally more difficult to solve than the model checking task.

The so-called *tableaux method* provides a systematic way to solve the model construction task. It was introduced for propositional logic by Beth in the 1950s [Bet55] and was refined and simplified by Smullyan [Smu95]. Fitting was the first to systematically extend it to the basic monomodal logics. More recently the tableaux method was extended to description logics [BS00]. Basically, the method goes as follows:

1. Start from a graph having a single root node that is labelled by the input formula A;
2. Build up a labelled graph by repeatedly applying the truth conditions, until none of the truth conditions applies any more;
3. Check whether the labelled graph can be transformed into a Kripke model M such that A is true at the root node of M.

In the last step, graph nodes are transformed into possible worlds, edges into accessibility relations, and node labels into valuations.

LoTREC was designed as a general and flexible tool to implement tableaux methods. The construction of the labelled graph starts from the input formula and applies so-called *tableau rules* which step by step break up the initial formula into subformulas, create new nodes, and add edges between nodes. We call the graphs that are produced during this process *premodels*. The construction of such premodels is done systematically in LoTREC. The iterated application of the rules leads to a *saturated premodel*: a premodel where no rule can be applied any more.

O. Gasquet et al., *Kripke's Worlds*, Studies in Universal Logic,
DOI 10.1007/978-3-7643-8504-0_3, © Springer Basel AG 2014

In this chapter we solve the problem of finding a model for a given formula in the class of *all* Kripke models. This is the model finding problem in modal logic **K**. In Chaps. 4, 5, and 7 we will explain how to proceed if we want the model to satisfy some constraints.

We start by giving a formal definition of labelled graphs (Sect. 3.1). We then illustrate in an example how the tableaux method works: we build a model for a given formula (Sect. 3.2). Generalising this we discuss how it can be done in a systematic way (Sect. 3.3). Then we explain the terminology of tableaux systems (Sect. 3.4) and introduce the part of the language of LoTREC allowing us to describe tableau rules (Sect. 3.5). After that, we show how to implement the method in LoTREC (Sect. 3.6), and we give the full set of rules for modal logic **K** (Sect. 3.7). We extend this method to deal with multiple modalities, i.e., with the multimodal logic $\mathbf{K}_n$ (Sect. 3.8) and the description logic **ALC**, which is a variant of $\mathbf{K}_n$ (Sect. 3.9). After presenting some optimisations (Sect. 3.10) we finally introduce the formal properties of soundness, completeness and termination of a tableaux system (Sect. 3.11).

3.1 Definition of Labelled Graphs

We start by a definition of structures that are a bit more liberal than Kripke models: while in the latter nodes can only be labelled by elements of the set of atomic formulas $\mathcal{P}$, we now allow nodes to be labelled by complex formulas. Moreover, we do not distinguish node labels and edge labels: there is a unique set of labels `L`. We call such structures *labelled graphs*.

Definition 8 Given a set of labels `L`, a labelled graph is a tuple $\mathtt{M} = (\mathtt{W}, \mathtt{R}, \mathtt{V})$ where:

- `W` is a non-empty set;
- $\mathtt{R}\colon \mathtt{L} \longrightarrow 2^{\mathtt{W}\times\mathtt{W}}$;
- $\mathtt{V}\colon \mathtt{L} \longrightarrow 2^{\mathtt{W}}$.

`M` is *finite* if `W` is finite and `R` and `V` have value $\emptyset$ almost everywhere on `L`.

`W` is the set of *nodes*. `R` is an *edge labelling function* which associates to every label a binary relation on `W`. `V` is a *node labelling function* which associates to each label a set of nodes from `W`.

We recall our typewriter font convention of Sect. 1.8: we use different fonts in order to distinguish what we write on paper—such as M and w—from what we store on a computer, such as `M` and `w`. Therefore `M` and `w` are as stipulated in the above Definition 8, while M and w are as stipulated in Definition 1 of Sect. 1.6 (page 14).

Kripke models can be viewed as particular labelled graphs where all the labels are atomic: we may turn a model $M = (W, R, V)$ with node labels $\mathcal{P}$ and edge labels $\mathcal{I}$ into a labelled graph by first setting $\mathtt{L} = \mathcal{P} \cup \mathcal{I}$ and then letting `M` be the graph $\mathtt{M} = (\mathtt{W}, \mathtt{R}, \mathtt{V})$ such that $\mathtt{W} = W$, $\mathtt{R} = R$, and $\mathtt{V} = V$. We then have that $\mathtt{V}(I) = \emptyset$ for every $I \in \mathcal{I}$ and $\mathtt{R}(P) = \emptyset$ for every $P \in \mathcal{P}$. The other way round, every labelled graph can be viewed as a model: it suffices to set $\mathcal{I} = \mathcal{P} = \mathtt{L}$.

3.2 Building an Example Model by Hand

We begin with the simplest version of the model finding problem: the set of edge labels is a singleton and the Kripke model that we want to build does not have to satisfy any particular constraint. This is the model construction problem in the monomodal logic **K**. Thus, instead of having multiple accessibility relations $R(I)$ and multiple modal connectives $[I]$ and $\langle I \rangle$, we only have one relation $R \subseteq W \times W$ that interprets the two modal connectives $\Box$ and $\Diamond$. Let us recall their truth conditions:

$$M, w \Vdash \Box A \quad \text{iff for every world } u, (w,u) \in R \text{ implies } M, u \Vdash A;$$
$$M, w \Vdash \Diamond A \quad \text{iff there exists a world } u \text{ such that } (w,u) \in R \text{ and } M, u \Vdash A.$$

So our language has the primitive connectives $\bot$, $\neg$, $\wedge$, $\vee$, $\Box$, and $\Diamond$. Under the LoTREC syntax declarations of Sect. 2.2 they are written in prefix notation as `False`, `not`, `and`, `or`, `nec`, and `pos` and they are displayed in infix notation as `FALSE`, `~`, `&`, `v`, `[]`, and `<>`. Our example formula is

$$A = \Box P \wedge \big(\Diamond Q \wedge \Diamond(R \vee \neg P)\big)$$

In LoTREC it is written

```
and nec P and pos Q pos or R not P
```

and is displayed as

```
[]P & <> Q & <>(R v ~P)
```

Our aim is to build a Kripke model $M = (W, R, V)$ such that $M, w_0 \Vdash A$ for some $w_0 \in W$.

3.2.1 *The Idea: Apply the Truth Conditions*

We are going to construct the Kripke model M step by step, through a series labelled graphs as defined in Definition 8. We start from an initial graph made up of a single node `w0` that is labelled by the formula A written in LoTREC's typewriter font, i.e. by the formula `A` = `[]P & <> Q & (<>R v ~P)`. (Remember that `&` is associative and that therefore no parentheses are displayed around the subformula `<> Q & (<>R v ~P)`, cf. Sect. 2.2.5.) This initial node `w0` (the root node) is intended to be the actual world of a model where the formula is true, provided we succeed in finding such a model. So this makes up an initial LoTREC graph

$$\mathtt{M}_0 = \big(\{\mathtt{w0}\}, \emptyset, \mathtt{V}_0\big)$$

such that $\mathtt{V}_0(\mathtt{A}) = \{\mathtt{w0}\}$ and $\mathtt{V}_0(\mathtt{B}) = \emptyset$ for all formulas `B` different from `A`.

During the transformation of M we identify "Formula A should *be true at* the world w of M" with "Node w of M is labelled by A." When the latter is the case we also say that A belongs to w.

Suppose a model M with a world w where the formula $A \wedge B$ is true. Formally this is written $M, w \Vdash A \wedge B$. According to the truth conditions, both A and B should also be true at w. Let M be the corresponding labelled graph and let w be the node associated to w. In order to respect the truth conditions we extend the node labelling function V in a way such that both A and B label w, i.e., such that w ∈ V(A) and w ∈ V(B).

During the model building process we made the 'working hypothesis' that there is a model of A. However, it may be the case that no model for A exists at all. Then at some point our hypothesis should become untenable. For instance, consider the formula $P \wedge \neg P$ and let us start by an initial graph M whose only node w0 is labelled by P & ~P. If we proceed as above then we have to add both P and ~P to w0. By that it becomes obvious that we will never be able to associate a Kripke model to M. It is for that reason that we do not call the labelled graph that results from the model construction procedure a model but only a *pre*model.

Observe that while the role of the truth conditions is to compute the truth value of a complex formula *bottom-up* (starting from atomic formulas P), the tableaux method proceeds *top-down*: it starts from a complex formula and decomposes it step by step according to the truth conditions. The structures built during that procedure are intermediate steps on the way to a Kripke model and therefore also have to store truth values for non-atomic formulas.

Once we have finished the construction of a premodel M, two tasks remain: first, we have to check whether M can be turned into a Kripke model M having the form as required by Definition 1 of Sect. 1.6; second, we have to check whether $M, w_0 \Vdash A$. The second task is nothing but model checking. The first will be explained in detail in Sect. 3.2.3.

3.2.2 Decomposing the Example Formula

Let us turn back to our example formula $A = \Box P \wedge (\Diamond Q \wedge \Diamond(R \vee \neg P))$ and let us examine in detail how its premodel is built step by step. As we have said, we hypothesise the existence of a model M and a world w_0 such that $M, w_0 \Vdash A$.

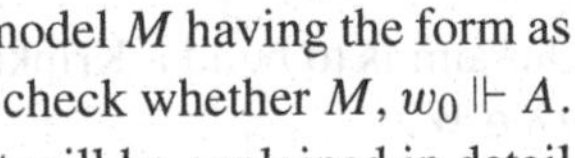

Fig. 3.1 Initial graph for $A = \Box P \wedge (\Diamond Q \wedge \Diamond(R \vee \neg P))$

[] P & <> Q & <> (R v ~ P)

w0

The Initial Graph We start with an initial premodel M = (W, R, V) where W = {w0}, R = ∅ and w0 is labelled by the set { [] P & (<> Q & <>(R v ~P)) }. In words, M has a single world w0 containing the input formula A, as shown in Fig. 3.1.[1]

Connective $\wedge$: Splitting Conjunctions According to the truth condition for conjunctions of Sect. 2.3, the hypothesis that $M, w_0 \Vdash \Box P \wedge (\Diamond Q \wedge \Diamond(R \vee \neg P))$ requires that:

- $M, w_0 \Vdash \Box P$, and
- $M, w_0 \Vdash \Diamond Q \wedge \Diamond(R \vee \neg P)$.

What we do is to simply add these two constraints to the node w0: we extend the node labelling function V of M so that it becomes

```
V(w0) = { [] P & (<> Q & <> (R v ~P)),
          [] P,
          <> Q & <> (R v ~P) }
```

By the same reasoning, the new assumption $M, w_0 \Vdash \Diamond Q \wedge \Diamond(R \vee \neg P)$ leads to a new premodel M with labelling function

```
V(w0) = { [] P & (<> Q & <> (R v ~P)),
          [] P,
          <> Q & <> (R v ~P),
          <> Q,
          <> (R v ~P) }
```

The results of these two steps are displayed in Fig. 3.2.

Connective $\Diamond$: Creating Successors According to the truth condition for the 'possible' operator, assuming that $\Diamond A$ is true in a world w imposes the existence of a world u such that u is accessible from w, i.e., such that $(w, u) \in R$, and such that A is true in u.

Consider the formula <> Q in the node w0 of Fig. 3.2. We should have a successor node, let it be u, that is accessible from w0, i.e., (w0, u) ∈ R, and that contains Q. Similarly, as w0 also contains <> (R v ~P), we should have a successor node v accessible from w0 containing R v ~P. Figure 3.3 depicts the premodel resulting from this step of the construction.

---

[1] We recall that although LoTREC does not display parentheses around <> Q & <>(R v ~P) because & is declared to be associative, we nevertheless did so here—just by declaring & to be non-associative in the language tab—in order to make clear the order of decomposition of A. That order is determined by the way we wrote down A in LoTREC's prefix notation, viz. as

```
and nec P and pos Q or pos R not P
```

as opposed to `and and nec P pos Q or pos R not P`.

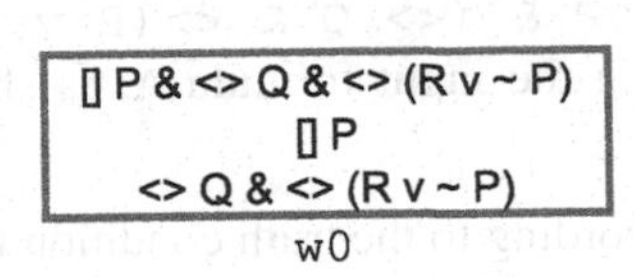

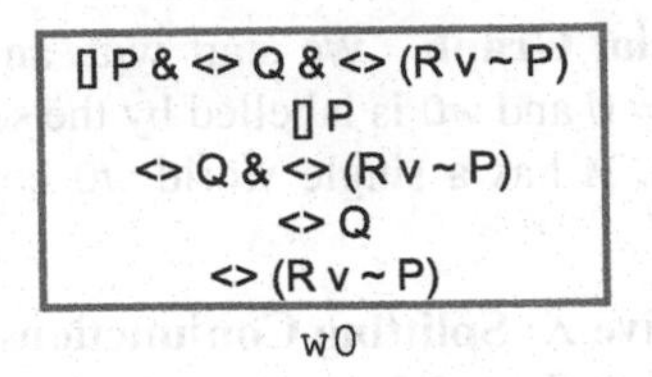

Fig. 3.2 Rule for $\wedge$ applied twice to the graph of Fig. 3.1

Fig. 3.3 Rule for $\Diamond$ applied to the graph of Fig. 3.2

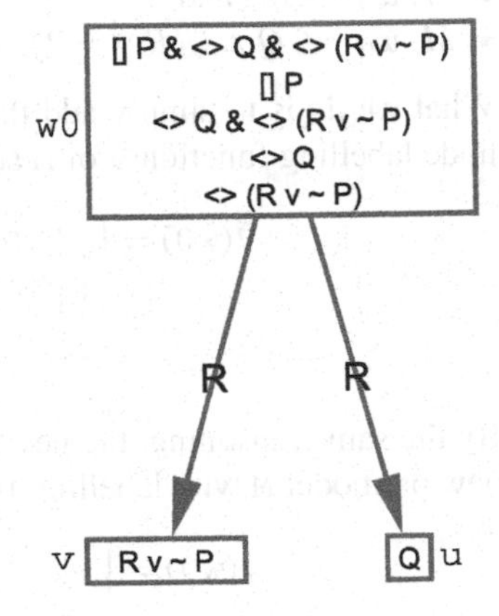

The reader may wonder why one should create two different possible worlds `u` and `v` for the above 'possible' operators. The answer is simple: since we are interested in maximizing our chances of finding a model, we would like to avoid contradictions as long as possible. It is therefore preferable to create one world per 'possible' formula `<> Ai`, so that the respective subformulas `Ai` will be separated in two different sets of node labels.

Connective $\Box$: Propagating Necessities Consider the other new assumption that we have introduced: $M, w_0 \Vdash \Box P$. According to the truth condition for the 'necessary' operator, $\Box P$ is true at w_0 iff P holds at every successor of w_0. That is why we extend the last premodel to a premodel where `u` is labelled by `Q` and `P` and where `v` is labelled by `R v ~P` and `P`. This is illustrated in Fig. 3.4.

Connective $\vee$: Duplicating the Premodel A disjunction is satisfied if at least one of its subformulas is. For instance, the formula `R v ~P` labelling `v` imposes that one of the formulas `R` or `~P` labels `v`.

What we do is to duplicate our premodel into two distinct candidate premodels: one containing `R` and the other containing `~P`. They are respectively depicted as `premodel.1` and `premodel.2` of Fig. 3.5.

Clashes: Introducing the Connective $\perp$ The connective $\perp$, displayed `FALSE` in LoTREC, is of arity 0 and cannot be decomposed. As soon as a graph has a node

Fig. 3.4 Rule for □ applied to the graph of Fig. 3.3

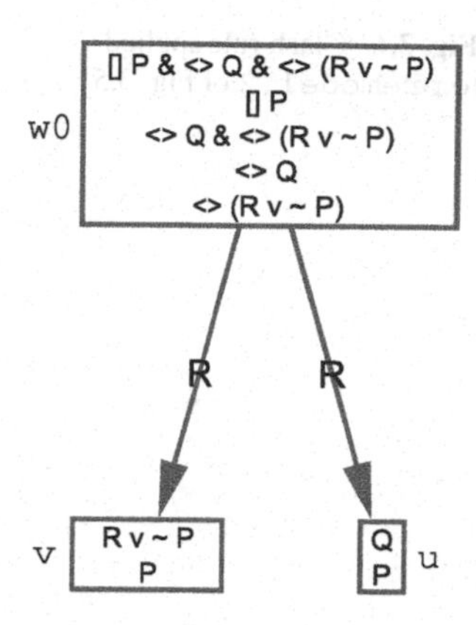

labelled FALSE we know that there is no chance to turn that graph into a legal Kripke model. When trying to build a model we therefore have to look out for this connective: it tells us that there is no hope of building a Kripke model from that graph.

The same is the case when a node contains both P and ~P, for some P: we say that there is a *clash*. This case can be reduced to the case of a node containing FALSE if we add a rule which adds FALSE to every node containing some P and ~P.

While there is no clash in premodel.1, there is one in premodel.2, as illustrated in Fig. 3.6. (As you can see, in order to signal contradictions LoTREC highlights nodes that are labelled FALSE.)

3.2.3 Extracting a Kripke Model from a Premodel

By now we have drawn all the consequences of the truth conditions, resulting in two premodels premodel.1 and premodel.2, and no further constraint can be

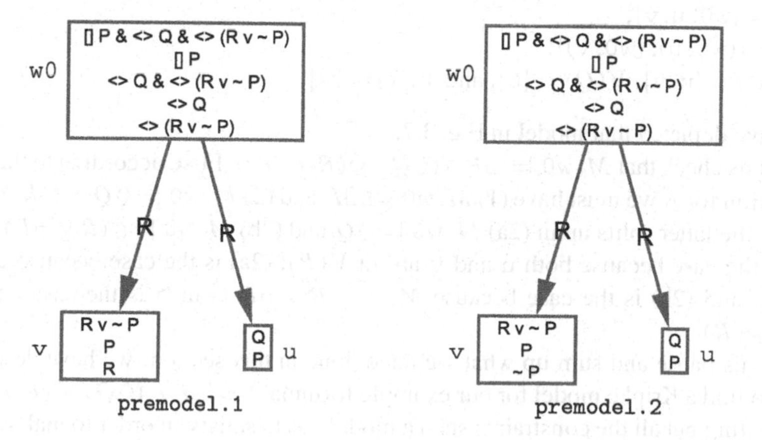

Fig. 3.5 Rule for ∨ applied to the graph of Fig. 3.4, creating premodel.1 and premodel.2

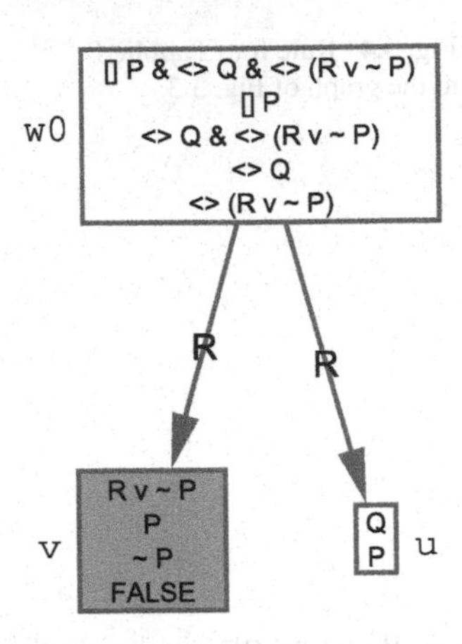

Fig. 3.6 Clash rule applied to `premodel.2` of Fig. 3.5

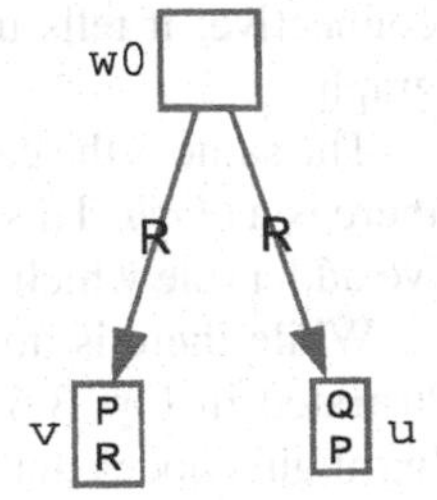

Fig. 3.7 The model M extracted from the premodel `premodel.1`

applied. Let us start by inspecting the latter: it cannot be transformed into a model because at v in Fig. 3.6, the formula FALSE is required to hold, which is impossible.

On the other hand, the premodel `premodel.1` can be transformed into a model, viz. the model $M = (W, R, V)$ such that:

- $W = \{\mathtt{w0}, \mathtt{u}, \mathtt{v}\}$;
- $R = \{(\mathtt{w0}, \mathtt{u}), (\mathtt{w0}, \mathtt{v})\}$;
- $V(P) = \{\mathtt{u}, \mathtt{v}\}$, $V(Q) = \{\mathtt{u}\}$, and $V(R) = \{\mathtt{v}\}$.

We have depicted that model in Fig. 3.7.

Let us check that $M, \mathtt{w0} \models \Box P \wedge (\Diamond Q \wedge \Diamond(R \vee \neg P))$. First, according to the truth condition for $\wedge$ we must have (1) $M, \mathtt{w0} \models \Box P$ and (2) $M, \mathtt{w0} \models \Diamond Q \wedge \Diamond(R \vee \neg P)$, where the latter splits up in (2a) $M, \mathtt{w0} \models \Diamond Q$ and (2b) $M, \mathtt{w0} \models \Diamond(R \vee \neg P)$. Now (1) is the case because both u and v are in $V(P)$; (2a) is the case because u is in $V(Q)$; and (2b) is the case because $M, \mathtt{v} \models R \vee \neg P$ (which is the case because $M, \mathtt{v} \models R$).

Let us pause and sum up what we have done in this section: we have described how to find a Kripke model for our example formula $A = \Box P \wedge (\Diamond Q \wedge \Diamond(R \vee \neg P))$ by spelling out all the constraints such a model has to satisfy in order to make A true at some node. We are now going to show how this can be done systematically.

3.3 How to Turn Truth Conditions into Tableau Rules

What we did in order to decompose our example formula was to apply the truth conditions for the connectives $\wedge$, $\Diamond$, $\Box$, and $\vee$. In the very end we also applied the truth conditions for $\neg$ and $\bot$ in order to decide whether the resulting premodel is indeed a Kripke model. This allowed us to build a Kripke model for our example formula. In order to turn this into a procedure that applies to *every* formula of the modal language we need a systematic way of breaking up formulas into subformulas: we should be able to decompose any formula in a way such that we always end up with a graph where we can decide whether it is a model by checking whether the clash rule applies.

There is however something missing: the rules we have seen so far do not allow us to decompose negations of complex formulas, i.e. formulas of the form $\neg\neg A$, $\neg(A \wedge B)$, $\neg(A \wedge B)$, $\neg\Diamond B$, and $\neg\Box B$. Take $\neg\neg A$: the truth condition for $\neg$ tells us *not* to add $\neg A$ to a node containing it. Doing things in this way would lead to configurations of positive and negative constraints that would be rather complex to manage.

Fortunately things can be done in a simpler way. The truth condition for negation tells us that $M, w \Vdash \neg\neg A$ if and only if $M, w \Vdash A$. Hence whenever a node contains $\neg\neg A$ we can add A to that node and proceed. Similarly, whenever a node contains $\neg(A \wedge B)$ we may exploit that $\neg(A \wedge B) \leftrightarrow \neg A \vee \neg B$ is true at every world of every Kripke model: we therefore can apply the rule for $\vee$ and copy the current premodel, add $\neg A$ to one and $\neg B$ to the other and proceed. Actually such equivalences exist for all combinations of $\neg$ with the connectives. Here is the entire list:

$$\neg\neg A \leftrightarrow A$$

$$\neg(A \wedge B) \leftrightarrow \neg A \vee \neg B$$

$$\neg(A \vee B) \leftrightarrow \neg A \wedge \neg B$$

$$\neg\Diamond A \leftrightarrow \Box\neg A$$

$$\neg\Box A \leftrightarrow \Diamond\neg A$$

For example when a node contains `~[] A` we act as if it contained `<> ~A`: we create a new node, link it, and add `~A` to its labels.

Observe that the last two rules reflect the duality of necessity and possibility, as manifested by the valid equivalences $\neg\Box\neg A \leftrightarrow \Diamond A$ and $\neg\Diamond\neg A \leftrightarrow \Box A$ that we have already mentioned in the end of Sect. 2.3.

Finally, we observe that $M, w \Vdash \neg\bot$ is the case for every world and every model. A node label `~FALSE` is therefore a constraint that is satisfied right away: we can safely forget it.

Remark 9 One might apply the above equivalences as much as possible, replacing the left-hand side by the right-hand side. The result will then be a formula in *negation normal form*: negation symbols only occur immediately in front of atomic formulas. For example, the negation normal form of the formula $\neg(P \wedge \neg(Q \wedge \neg R))$ is $\neg P \vee (Q \wedge \neg R)$.

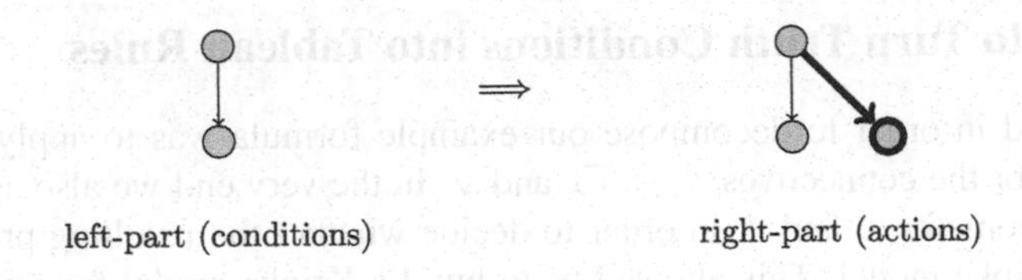

Fig. 3.8 Rule schema

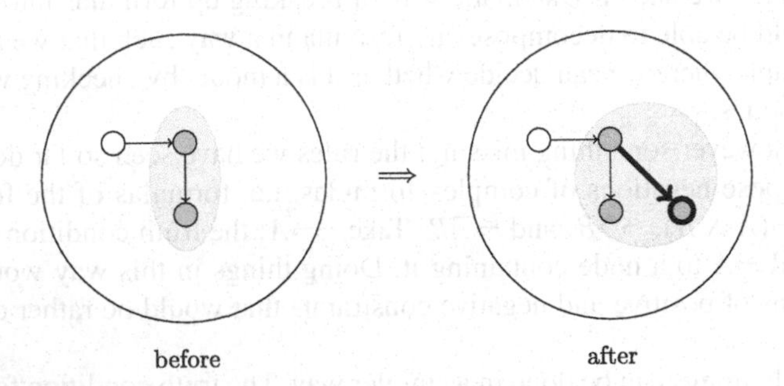

Fig. 3.9 Rule application

3.4 Tableaux: Some Fundamental Notions

In the introduction of this chapter we have already mentioned that our model construction method generalises the tableaux methods that are familiar from the literature. In particular, instead of trees we work with what we called premodels, which are nothing but labelled graphs. In this section we list and explain the most important concepts of our model-building approach.

Tableau Rules We decompose formulas by means of *rules*. What these rules do is to add nodes, edges, node labels, and edge labels to graphs if some conditions are satisfied.

Rules can be defined as a pair of graphs: a left-hand side graph and a right-hand side graph. This is illustrated in Fig. 3.8. When the left-hand graph matches a subgraph of a given premodel then the rule is *applicable* to that subgraph. The *application* of the rule consists in replacing this subgraph by (the instance of) the right-hand side graph. This is illustrated in Fig. 3.9.

LoTREC rules are restricted in an important way: they only *add* items—nodes, edges and labels for nodes and edges—to the current premodel and they never delete anything. We call such rules *monotonic*.

Thanks to our monotonicity constraint a LoTREC rule can be described by a couple that is made up of the following:

- A *set of conditions* describing the left-hand side graph: when the conditions are all true then the rule is applicable;
- A *list of actions* specifying the various items to be added to the current premodel in order to obtain the next premodel, together with the places where they have to

be added. These items are nodes, edges, node labels, and edge labels. They are added in the order the respective actions appear in the list. A node or an edge has to be created before it can be labelled.

The conditions and actions existing in LoTREC are explained in more detail in Sect. 3.6 as well as in the later chapters.

Variables vs. Constants A rule typically concerns *every* formula of such or such form. For example, the rule for conjunction applies to every formula of the form $A \wedge B$. This means that A and B are supposed to stand for *any* formula. So the formula $\Box P \wedge (\Diamond Q \wedge \Diamond(R \vee \neg P))$ is of the form $A \wedge B$ because we may consider that A stands for $\Box P$ and B stands for $\Diamond Q \wedge \Diamond(R \vee \neg P)$. The symbols A and B are called *formula variables*.

Beyond formula variables, a tableau rule may also contain variables for edges. We shall make use of this in the rules for the multimodal version $\mathbf{K}_n$ of logic **K** in Sect. 3.8. We use the term *label variables* in order to talk about both node and edge variables.

In order to apply the conjunction rule to particular formulas we must know what A and B stand for: we have to ***substitute*** variables by formulas. Mathematically speaking, a substitution is a function from the set of variable symbols to the set of formulas. Of course, if a variable symbol occurs several times in a rule then each occurrence should be substituted by the same formula.

As we have already said, all the node names occurring in a tableau rule are variables, while variables other than world names have to be identified as such by the LoTREC keyword variable preceding it.

Sets of Premodels vs. Tableau Branches In the standard presentation of the tableaux method, a tableau is a set whose elements are called branches. We did not take over that terminology because it sounds a bit odd to use the term 'branch' for something that actually is a graph. We have decided to replace it by the term *premodel*: a tableau is a set of premodels. In a given set of premodels, each premodel other than the initial premodel has been created due to some disjunction. We will encounter other rules creating premodel copies in the tableaux procedures for the logics **LTL** and **PDL** in Chap. 7.

A tableaux procedure may proceed in two different ways:

- It may work depth-first and focus on the first premodel, postponing the construction of the other premodels that is only undertaken when the first attempt fails (this therefore requires backtracking);
- It may work breadth-first and build all the premodels simultaneously.

LoTREC goes the second way and explores premodels simultaneously and systematically: at each application of the disjunction rule it duplicates the current premodel, creating a new premodel copy. The original premodel and its copy share the part that was developed before the duplication.

Partial vs. Saturated Premodels A premodel M is *saturated* when every applicable rule has been applied; otherwise M is called *partial*. When every premodel is saturated then the tableaux procedure method stops. (There is a way of stopping the construction earlier, which will be explained in Sect. 3.6.2.)

Open vs. Closed Premodels Premodels with a node containing FALSE are called *closed*; otherwise they are called *open*.

A closed premodel cannot be turned into a legal Kripke model: this would require a possible world where $\bot$ is true, which cannot be the case. An open premodel that is saturated can directly be turned into a model: it suffices to drop the complex node labels, just as we did in our example in Sect. 3.2.3. In contrast, when an open premodel is partial then it may happen that a further application of some tableau rule adds FALSE to some node of the graph and therefore causes the premodel to be closed.

So the transformation of an open saturated premodel into a model is relatively simple in the case of the class of all Kripke models, i.e., in the case of model construction for modal logic **K**. The transformation requires a bit more work if we want the resulting Kripke model to satisfy some constraints, as we shall see in Chaps. 4, 5, and 7.

Tableaux Procedure A *tableaux procedure* is made up of a set of tableau rules together with one or more *strategies* applying these rules in some order. We are going to define a very simple language of strategies in Sect. 3.6.

3.5 The Language of Tableau Rules in LoTREC

In Sect. 3.2 we decomposed an example formula by means of rules, one rule per connective. In Sect. 3.3 we motivated why this should be done in a systematic way: for each connective, either there should exist a tableau rule for that connective, or—in the case of negation—there should be a tableau rule for each combination of that connective with the other connectives. We are now going to define LoTREC rules for each connective.

Tableau Rule for $\wedge$: Splitting Conjunctions In order to analyse formulas of the form A & B labelling some node w we have to add both A and B to w. This is achieved by a rule with the following condition and action parts:

- condition: node w contains a formula of the form A & B;
- actions:
  - add A to node w;
  - add B to node w.

We use the primitive hasElement in order to write the condition and write: hasElement w and variable A variable B. Observe that we use here the same prefix notation as introduced in Sect. 2.2. Observe also that A and B were prefixed by the keyword variable: when we write variable A then the symbol A is declared to be a variable. Variables are substituted by labels when a rule is applied. If the keyword variable is missing then LoTREC interprets A as a constant.[2]

In order to add the formulas A and B to node w we use the LoTREC keyword add that we have already introduced in Sect. 1.7 (page 16). There, the argument of add was a constant, i.e., a particular atomic formula. Here it is rather a variable, and we add the formulas A and B to node w by means of the two actions add w variable A and add w variable B. (Instead of "the formula A" we should actually rather say "the label the variable A stands for": when the rule is applied we suppose that A has been substituted in the condition part of the rule.) If ever the variable has not been substituted in the condition part then LoTREC does not know what to add and signals an error.

Here is the entire definition of the rule for conjunction in LoTREC:

```
Rule And
  hasElement w and variable A variable B

  add w variable A
  add w variable B
```

As one can see, a LoTREC rule has a rule name in the first line—And—, and it has a condition part and an action part. The rule name is arbitrary: we could as well have written and, conjunction, RuleForConjunction, etc.

Tableau Rule for ◇: Creating Successors When the main connective is ◇ then we have to create a successor of the current node and put the subformula following ◇ in that node. Successor creation is therefore a rule with the following conditions and actions:

- condition: a node w contains a formula of the form <> A;
- actions:
  - add some new node u;
  - link node w to u by an edge labelled R;
  - add A to u.

In LoTREC, we write this as follows:

```
Rule Pos
  hasElement w pos variable A
```

[2] We have already mentioned in Sect. 3.4 that the node name w should in principle also be declared to be a variable. However, there are no constant symbols denoting particular nodes in LoTREC. We therefore write w instead of variable w: by convention, the first argument of the 'add' action is always a variable.

```
createNewNode u
link w u R
add u variable A
```

The LoTREC action keywords createNewNode and link speak for themselves. We have already encountered them in Sect. 1.7 and we just recall that the createNewNode action creates a new graph node and associates it with the variable u.

The edge label R is a constant because it is neither a possible world nor prefixed by the keyword variable. As we shall see in Sect. 3.8 where we treat the multimodal version $\mathbf{K}_n$ of logic $\mathbf{K}$, one is also allowed to put an edge variable in that position. Remember that the node labels w and u are variables, even if we did not prefix them by the keyword variable.

Actions are executed in sequence. It is crucial that createNewNode u occurs before link w u R: if they were permuted then an error would occur. In contrast, the actions link w u R and add u variable A can safely be permuted.

Tableau Rule for □: Propagating Necessities When the main connective is □ then we propagate the argument of □ to all those nodes that are linked to the current node by the accessibility relation. This can be expressed by means of two conditions: we require a node w containing a formula of the form [] A and a link from w to some u. In other words, we have:

- conditions:
  - a node w contains a formula of the form [] A;
  - w is linked to a node u;
- action: add A to u.

We check that node w is linked to node u by the relation R by means of the LoTREC keyword isLinked and write isLinked w u R. The rule for □ is therefore:

```
Rule Nec
  hasElement w nec variable A
  isLinked w u R

  add u variable A
```

Tableau Rule for ∨: Duplicating the Graph The case of disjunction ∨ is special: we have to make a copy of the current premodel. Here is the description of the rule:

- condition: a node w contains a formula of the form A v B;
- actions:
  - duplicate the current premodel;
  - in one, add A to w;

– in the other one, add B to w.

Duplication is achieved in LoTREC by means of the keyword duplicate. The action duplicate premodel_copy creates a copy of the current premodel that is denoted by premodel_copy. Instead of the word premodel_copy we may give any other name to that copy. We can access node w of premodel_copy by writing premodel_copy.w. Accessing the node w of the current premodel is still possible by simply writing w.

We want to add A to the node w of the current premodel and we want to add B to the node w in its copy. We therefore write:

```
Rule Or
  hasElement w or variable A variable B

  duplicate premodel_copy
  add w variable A
  add premodel_copy.w variable B
```

Tableau Rule for Clashes If a node w is labelled both by A and its negation then we have a manifest contradiction, also called a clash. In that case we add FALSE—the LoTREC keyword for falsum—to w. This is achieved by the following rule:

```
Rule Clash
  hasElement w variable A
  hasElement w not variable A

  add w False
```

Note that the formula A stands for need not be atomic. We recall that False is the only predefined connective of LoTREC and that it is displayed as FALSE.

The Tableau Rules for ¬: Checking the Next Connective The set of rules defined so far suffices to analyse the example formula of Sect. 3.2.2. It however does not allow decomposition of formulas of the form $\neg A$. As we have explained in Sect. 3.3 we have to check the main connective of A:

- if A is of the form $\neg B$ then add B to w;
- if A is of the form $B \wedge C$ then proceed as if $\neg A$ was $\neg B \vee \neg C$;
- if A is of the form $B \vee C$ then proceed as if $\neg A$ was $\neg B \wedge \neg C$;
- if A is of the form $\Box B$ then proceed as if $\neg A$ was $\Diamond \neg B$;
- if A is of the form $\Diamond B$ then proceed as if $\neg A$ was $\Box \neg B$;
- if A is of the form $\bot$ then do nothing.

For example when ~[] B labels w then we proceed as if <> ~B was labelling w: we extend the premodel by adding a new successor node u that is accessible from w by an edge labelled R and that contains ~B.

In the language of LoTREC we formulate the following rules that should by now be self-explanatory.

```
Rule NotNot
  hasElement w not not variable A

  add w variable A

Rule NotOr
  hasElement w not or variable A variable B

  add w not variable A
  add w not variable B

Rule NotAnd
  hasElement w not and variable A variable B

  duplicate premodel_copy
  add w not variable A
  add premodel_copy.w not variable B

Rule NotPos
  hasElement w not pos variable A
  isLinked w u R

  add u not variable A

Rule NotNec
  hasElement w not nec variable A

  createNewNode u
  link w u R
  add u not variable A
```

A Constraint on the Form of LoTREC Rules The pattern-matching mechanism of LoTREC imposes two important constraints that every rule has to satisfy in order to be applicable.

Definition 9 A LoTREC rule ρ is called *legal* if:

1. The condition part of ρ contains one of the conditions among
   - `isNewNode w`
   - `hasElement w B`
   - `hasNotElement w B`
   - `isLinked w u B`

   for some (possibly complex) labels B and some node variables w and u;
2. Every variable *`variable`* `A` occurring in ρ also occurs in a condition having one of the above four forms.

All the rules that we have seen up to now satisfy the first constraint, simply because the LoTREC condition keywords we have introduced so far are exactly those of the first constraint. The simplest case of a rule violating the first constraint is a rule with an empty condition part. There will be more illegal rules: from Chap. 5 on we are going to introduce further condition keywords that are not in the above list, such as isMarked. So an example of a rule whose condition part violates the first constraint is:

```
Rule InapplicableRule
  isMarked w variable A

  add w variable A
```

That rule will never be triggered: it is just as if it did not exist.

An example of a rule that satisfies the first constraint but violates the second is the following:

```
Rule InapplicableRule
  hasElement w variable A
  isMarked w variable B

  add w variable B
```

Indeed, although variable B occurs in the condition part, it does not occur in the condition hasElement w variable~A. Again, that rule will never be applied by LoTREC, under whatever strategy.

Remember that node variables are not preceded by the keyword variable. Therefore the second constraint does not concern them. This is as it should be: in a rule with the action createNewNode w we do not expect the node variable w to occur in the condition part of the rule.

3.6 Strategies: How to Combine LoTREC Rules

In Sect. 3.2 we built a model for an example formula by hand, applying the tableau rules in the appropriate order. In this section we are going to see how this can be done automatically with LoTREC. Figure 3.10 depicts the architecture of LoTREC in terms of an input formula, a logic definition, and a graphical output (that might be transformed into a model or not). The most delicate part of a logic definition is that of a *strategy*. We first give a strategy for the example formula and then define a strategy working for every input formula.

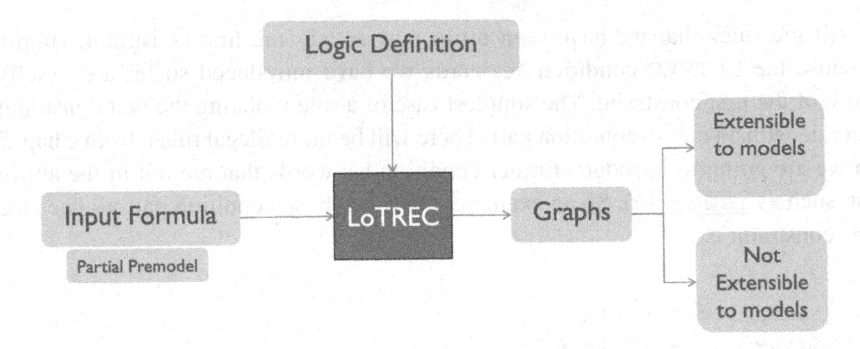

Fig. 3.10 The architecture of LoTREC

3.6.1 Rule Application in LoTREC: Everywhere and Simultaneously

Let us start by implementing in LoTREC the step-by-step construction of Sect. 3.2 for the formula $A = \Box P \land (\Diamond Q \land \Diamond(R \lor \neg P))$. What we did was to apply the tableau rules in a particular order: first we applied the `And` rule twice, then the `Pos` rule, etc. So we have to tell LoTREC to first apply rule `And`, then reapply it, then apply rule `Pos`, then rule `Nec`, and finally rule `Or`. This is nothing but a particular strategy. It can be written straightforwardly in LoTREC:

```
Strategy Strategy_For_The_Example
  And
  And
  Pos
  Nec
  Or
  Clash
end
```

Exercise 20 In LoTREC's strategy tab, enter `Strategy_For_The_Example`. Then enter the example formula `A = and nec P and pos Q pos or R not P` in the formula field and press the "Build Premodels" button.

You may use LoTREC's step-by-step mode in order to execute the rules one-by-one. You can define breakpoints by ticking the appropriate rules in the list that pops up when calling that mode.

When a rule is called by the strategy then it is applied *simultaneously* and *wherever possible*: all possibilities of applying it are exploited in parallel. In the situation depicted in Fig. 3.4 it therefore suffices to call the rule `Nec` just *once* in the strategy in order to propagate the subformula `P` of `[] P` at `w0` to *both successors* of `w0`, i.e., to `u` and `v`. (The reader should note that a single simultaneous application of the And rule is not enough in order to decompose our example formula because the conjunction in the subformula and pos Q pos or R not P is nested: it only matches the condition part of the And rule once the original formula A has been decomposed by a previous application of the And rule.) The application is also *unique*: LoTREC tries to apply a rule only once, namely when the strategy calls the Nec rule. In our strategy above we therefore had to say explicitly that the And rule is applied twice in sequence.

Once the first occurrence of the And rule in Strategy_For_The_Example has been applied, the input formula A is *no longer active*: although we did not remove it, the second occurrence of the And rule does not apply to it: its subformulas [] P and <> Q & <> (R v ~ P) are not added to the node yet again. In fact, LoTREC exploits each possibility of applying a rule only once. Of course, if the And rule is called again by the strategy then LoTREC will try again to apply it wherever possible.

Rule order is important. Suppose we permute the second And rule with the Pos rule. Starting with the same formula A, the premodel that one would obtain with the modified strategy would result in a single node containing the original formula A and its subformulas [] P, <> Q & <> (R v ~ P), <> Q, and <> (R v ~P): the first And rule splits the first & when called, then rule Pos is called but does not apply because there is no <> formula yet, i.e., there is no formula whose main connective is <>. Then the second And applies and splits the second &, and the computation stops here because the Nec rule does not apply... while it is *now* that rule Pos would be useful!

Exercise 21 Apply the second strategy by hand to the example formula A. Then do so with LoTREC, using the step-by-step mode to check whether your steps on paper match LoTREC's.

3.6.2 Stopping and Killing Strategies

The only connective to which we did not associate a tableau rule is the falsum symbol $\bot$. This is in order: as soon as a premodel contains a node w labelled FALSE, we know right away that we will never be able to turn the premodel into a model satisfying all the formulas labelling w, whatever further rules we apply. (Remember that rules never delete anything.)

As it is useless to perform any further actions, we may as well *block* any further rule application to premodels containing FALSE.

The stop action asks LoTREC to stop the development of all those premodels having some node labelled FALSE:

```
Rule Stop
  hasElement w False

  stop w
```

The `Stop` rule improves the performance of strategies: it typically prevents superfluous rule applications once a clash has been detected. In the case of many unsatisfiable formulas, this allows us to avoid superfluous computations and by that to conclude at a much earlier stage that the input formula has no model.

When we try to build a model we are often not interested in closed premodels. Beyond stopping their development we may even entirely drop them from the set of current premodels. This can be done by means of the following self-explaining rule:

```
Rule Kill
  hasElement w False

  kill w
```

Exercise 22 Insert the `Kill` rule at the end of `Strategy_For_The_Example` and observe how LoTREC's behaviour differs. (Check the tab listing the resulting set of premodels.)

3.6.3 A Strategy for Any Formula: Saturation by Rule Iteration

The rules that we have introduced allow decomposition of other formulas beyond the example formula `A`. It may then of course happen that one has to apply the `Pos` rule twice—think of a formula such as `<> <> P`.

Exercise 23 Write strategies that build models for the formulas $\Diamond P$, $\Diamond\Diamond Q$, and $\Diamond\Diamond\Diamond R$. Try to write a single strategy that works for all three formulas. Do the same for the formulas $\Diamond(P \land Q)$ and $R \land \Diamond(S \land P)$.

Is there a unique strategy that can completely decompose every formula? In fact, all rules should be applied as many times as needed. For example, one has to apply the `Pos` rule several times: for $\Diamond\Diamond\Diamond P$ you have to apply it three times, for $\Diamond\Diamond\Diamond\Diamond P$ you have to apply it four times, and so on. We therefore have to tell LoTREC to repeat the application of the `Pos` rule *as much as possible*. To do this, LoTREC provides the keyword `repeat`. Here is a strategy allowing decomposition of any formula that is only built with the connective $\Diamond$:

```
Strategy Strategy_For_Multiple_Pos
  repeat
    Pos
  end
end
```

The keyword end allows LoTREC to identify the sequence of rules to be repeated. The repeat loop stops when there is no more ◊ formula to be decomposed.

One may repeat more than one rule by writing the rule names between the keywords repeat and end. The following strategy lists all rules we have seen so far:

```
Strategy K_Strategy
  repeat
    Stop
    Clash
    NotNot
    And
    NotOr
    Or
    NotAnd
    Pos
    NotNec
    Nec
    NotPos
  end
end
```

The strategy repeatedly tries to apply each tableau rule in sequence, until none of them applies any more.

Exercise 24 Run K_Strategy on the formula $P \wedge \neg P \wedge \Diamond Q$. Then drop the Stop rule from K_Strategy (by editing the strategy and erasing it). What is the difference between the premodels returned by LoTREC?

Observe that we might put the rules in a different order. The reader may check by executing the different versions in LoTREC's step-by-step mode that this does not change the outcome but only the way it is produced.

When LoTREC applies K_Strategy then it executes a particular program. That program combines the execution of the tableau rules by means of two program constructions: sequential composition by the mere juxtaposition of tableau rules and iteration by the repeat loop. Both constructions are omnipresent in programming languages. Readers who have already written and run computer programs know that repeat loops sometimes do not terminate and can run forever. In Sect. 4.9 we will address that problem and show that while there exist indeed non-terminating LoTREC strategies, the above K_Strategy always terminates. So it will never happen that the rules in the repeat loop of K_Strategy are applied infinitely many times.

The strategy K_Strategy is *fair*: it loops through all rules, and as each rule application terminates we are sure that every applicable rule gets its chance and will eventually be applied.

Exercise 25 Run K_Strategy on $\Box P \wedge \Diamond\neg P$ and $\Box P \wedge \Box\neg P$. What do you conclude about the satisfiability of these formulas?

3.6.4 Nesting Strategies

In LoTREC, strategies may call other strategies. For instance, let us group the rules for the boolean connectives in the following strategy.

```
Strategy BooleanRules
  Stop
  Clash
  NotNot
  And
  NotOr
  Or
  NotAnd
end
```

We can then modify `K_Strategy` and call `BooleanRules` inside the `repeat` loop:

```
Strategy K_Strategy_booleanMacro
  repeat
    BooleanRules
    Pos
    NotNec
    Nec
    NotPos
  end
end
```

As the reader may observe by implementing the above strategy, it behaves exactly as the original `K_Strategy`.

Inside a repeat-loop we may also call another repeat-loop. Here is an example.

```
Strategy K_Strategy_nested
  repeat
    repeat
      BooleanRules
    end
    Pos
    NotNec
    Nec
    NotPos
  end
end
```

This strategy applies the rules for the boolean connectives as long as possible before creating new nodes.

Exercise 26 Consider the sequence of formulas

$$\Diamond Q \wedge (P_1 \wedge \neg P_1)$$

$$\Diamond\Diamond Q \wedge \big((P_1 \wedge \neg P_1) \wedge P_2\big)$$

$$\Diamond\Diamond\Diamond Q \wedge \big(((P_1 \wedge \neg P_1) \wedge P_2\big) \wedge P_3\big)$$

$$\ldots$$

Observe that the contradiction is nested deeper and deeper in the formula. Run the strategies `K_Strategy_booleanMacro` and `K_Strategy_nested` on these formulas. Compare the depth of the premodels generated by LoTREC in both cases.[3]

3.6.5 Prioritising Rule Application

As Exercise 26 illustrates, the size of the premodels (and with that LoTREC's execution time) may vary depending on the strategy. It is generally a good idea to try the `Stop` rule whenever possible: it avoids superfluous computation when the premodel under construction is already closed. Similarly, it is a good idea to apply the `NotNot`, `And`, and `NotOr` rules whenever possible: they break down formulas whenever possible, while not duplicating premodels. The `Or` and `NotAnd` rules do the latter and therefore increase the amount of memory LoTREC needs. They are therefore best put after the `NotNot`, `And`, and `NotOr` rules.

All these considerations lead us to improve the strategy for classical propositional logic by executing the above rules in the corresponding order. Here is a way of doing this:

```
Strategy CPL_Strategy_prioritised
  repeat
    repeat
      repeat
        Stop
        Clash
      end
      NotNot
      And
      NotOr
    end
  Or
```

[3] ToDo: A verifier que c'est bien different (y avait une erreur trouvee par Teddy).

```
  NotAnd
  end
end
```

The above nesting of repeat loops looks a bit clumsy. This can be avoided by using the LoTREC keyword firstRule. The execution of the block

firstRule Rule_1 Rule_2 ... Rule_n end

applies the first applicable rule of the list Rule_1 Rule_2 ... Rule_n. So LoTREC starts by checking whether Rule_1 is applicable: if so then only Rule_1 is executed and LoTREC stops; if not then it is checked whether Rule_2 is applicable: if so then only Rule_2 is executed and LoTREC stops; and so on. If no rule is applicable then the execution does nothing. The strategy BooleanRules can therefore be improved as follows:

```
Strategy BooleanRules_with_firstRule
  firstRule
    Stop
    Clash
    NotNot
    And
    NotOr
    Or
    NotAnd
  end
end
```

Exercise 27 Rewrite the above strategy CPL_Strategy_prioritised using the firstRule keyword.

Sometimes, inside a loop such as repeat firstRule R_1 R_2 R_3 R_4 end end, one wishes to treat the rules of the subsequence R_2 R_3 on a par. This can be achieved in two different ways. The first is to define a substrategy:

```
Strategy SomeSubstrategy
  firstRule
    R_2
    R_3
  end
end
```

and to call it in the loop:

```
Strategy ExampleStrategy
  repeat
    R_1
```

```
    SomeSubstrategy
    R_4
  end
end
```

Alternatively, one may use the LoTREC keyword and write:

```
Strategy ExampleStrategy
  repeat
    R_1
    allRules
      R_2
      R_3
    end
    R_4
  end
end
```

The predefined logics in LoTREC are all written in a way that avoids the allRules keyword. The reader may however use it whenever it is convenient.

3.7 Exercises: Adding Connectives

Exercise 28 (Other boolean connectives as abbreviations) In Sect. 2.1 we have defined the connectives $\rightarrow$ and $\leftrightarrow$ as abbreviations. In LoTREC, in line with our previous notations we call them `imp` and `equiv` and we display them as -> and <->.

1. Write a LoTREC rule for the connective `imp` that adds $\neg A \vee B$ to nodes containing $A \rightarrow B$.
2. Write a LoTREC rule for `equiv` that adds $(A \rightarrow B) \wedge (B \rightarrow A)$ to nodes containing $A \leftrightarrow B$.
3. Observe that you also need rules for the negations of these connectives, cf. Sect. 3.3. Write them down. (Call them `NotImp` and `NotEquiv`.)
4. Add these rules to `K_Strategy`. Run LoTREC on `P <-> (Q <-> R)`.
5. Do the same for the exclusive 'or' $\oplus$.

Exercise 29 (Other primitive boolean connectives) Alternatively, we may consider implication and equivalence to be primitive connectives. In that case we have to decompose them directly.

1. Write LoTREC rules for `imp` and `equiv` that match the truth conditions for $\rightarrow$ and $\leftrightarrow$. Don't forget the rules for the negated connectives.
2. Add these rules to `K_Strategy`. Run LoTREC on `P <-> (Q <-> R)`. Observe that fewer node labels are added than in the previous exercise.

3.8 From Monomodal K to Multimodal Logic $\mathbf{K}_n$

Up to now we only considered the monomodal logic **K**, where the set of edge labels $\mathcal{I}$ is a singleton. We now consider the more general case where $\mathcal{I}$ is an arbitrary set. We still suppose that the accessibility relations $R(I)$ do not have to satisfy any constraint. This is the model-finding problem in multimodal logic $\mathbf{K}_n$. It is not necessary to define the tableaux procedure for the multimodal $\mathbf{K}_n$ from scratch: it can be easily adapted from the model construction for the monomodal **K**.

As for the language, in $\mathbf{K}_n$ the modal operators are not $\Diamond$ and $\Box$ but parametrised operators of the form $\langle I \rangle$ and $[I]$, where $I \in \mathcal{I}$ is an edge label. Hence, to reuse the method of **K** for $\mathbf{K}_n$ we have to change the definition of the modal connectives. We take the definition of Table 2.2 in Sect. 2.2.

As for the rules, we can reuse the same set of rules defined above for monomodal **K**, except that the modal rules `Pos`, `Nec`, `NotPos` and `NotNec` have to be slightly modified: they should take the edge labels I into account. For example, the `Pos` rule is now defined as follows:

```
Rule Pos
  hasElement w posi variable I variable A

  createNewNode u
  link w u variable I
  add u variable A
```

So we now write `posi variable I variable A` instead of `pos variable A`. The variable `I` will be substituted by edge labels which are going to label the link that is created by the action `link w u variable I`.

Similarly, the `Nec` rule should be modified so that the `[I]`-formulas are only propagated along those edges that are labelled by `I`:

```
Rule Nec
  hasElement w nec variable I variable A
  isLinked w u variable I

  add u variable A
```

The modal rules `NotNec` and `NotPos` have to be modified similarly, by replacing each occurrence of `R` by `variable~I`.

As for the strategy, we can keep on using the strategy `K_Strategy` that we have defined for the monomodal logic **K** in Sect. 3.7 since the rule names are identical.

3.9 Description Logic ALC

The description logic **ALC** is basically a notational variant of the multimodal logic $\mathbf{K}_n$. However, the language of **ALC** has more constituents than that of $\mathbf{K}_n$:

names for graph nodes are part of the language. Such a language allows formulation of some more reasoning problems than those listed in Sect. 2.5. We here only present how the problem of concept satisfiability can be addressed in LoTREC.

We have already introduced the relevant connectives of **ALC** in Sect. 2.1.5. The LoTREC rules are below. We adopt here the standard notational conventions in description logics: a and b denote individuals, C denotes a concept, and R denotes a relation ('role').

```
Rule Pos
  hasElement a exists variable R variable C

  createNewNode b
  link a b variable R
  add b variable C

Rule Nec
  hasElement a forall variable R variable C
  isLinked a b variable R

  add b variable C
```

The reader may have noticed that these two rules can be obtained from the rules Pos and Nec of Sect. 3.8 by replacing pos by exists and nec by forall.

The rules for the cases $\neg\forall$ and $\neg\exists$ and for the boolean connectives $\sqcap$ and $\sqcup$ can be obtained in a straightforward way from those of propositional logic of Sect. 3.8. All our rules are therefore notational variants of those of Sect. 3.8 and we can keep the strategy for $\mathbf{K}_n$ of Sect. 3.8.

Exercise 30 Try to build models for the following **ALC** concepts:

$$\forall \text{parentOf.male} \sqcap \exists \text{parentOf.}\neg\text{male}$$

$$\forall \text{parentOf.male} \sqcap \forall \text{parentOf.}\neg\text{male}$$

Which of them is satisfiable? Why?

We have said in the preface that LoTREC cannot compete in performance with automated theorem provers that are specialised for particular logics. This is the case in particular for **ALC**: for example FaCT [Hor98b, Hor98a] is an optimised prover for **ALC** whose current version FaCT++ is among the fastest provers for **ALC** and some other description logics [TH06].

3.10 Taming the Rule for Disjunction

When formulas have a lot of disjunctions then K_strategy returns a lot of premodels. We now show how this can be improved by controlling the application of the rule for disjunction.

3.10.1 Redundant Disjunctions

Each application of the rule for disjunction generates a copy of the current premodel. This takes time and computer memory.

Premodel duplication can sometimes be avoided. Consider the formula $P \wedge (P \vee Q)$. Unsurprisingly, when run on `P & (P v Q)` the strategy `K_Strategy` of Sect. 3.6.3 outputs two saturated premodels: one consisting of a single node containing `P & (P v Q)`, `P`, `P v Q`; and another one consisting of a single node containing `P & (P v Q)`, `P`, `P v Q`, and `Q`. However, the second graph is redundant: it is clear that if we can build a model from the latter then we can also build a model from the former. Indeed, for every Kripke model M and every world w in M we have that $M, w \Vdash P \wedge (P \vee Q)$ if and only if $M, w \Vdash P$.

How can we block the application of the rule for disjunction to such redundant disjunctions? This can be achieved by adding two further conditions to the rule: neither of the disjuncts should be contained in the node. Here is the improved version of the `Or` rule:

```
Rule Or_withRedundancyCheck
  hasElement w or variable A variable B
  hasNotElement w variable A
  hasNotElement w variable B

  duplicate premodel_copy
  add w variable A
  add premodel_copy.w variable B
```

In our example the rule `Or_withRedundancyCheck` cannot be applied to `P v Q` because the node already contains `P`. Duplication of the premodel is thus avoided.

Exercise 31 Modify the rule `NotAnd` by inserting a redundancy check. (Call the resulting rule `NotAnd_withRedundancyCheck`.)

3.10.2 The Cut Rules

There is another kind of configuration where one can avoid unfruitful applications of the rule for disjunction. Consider node `v` of `premodel.2` in Fig. 3.5. It is obtained by applying the rule for disjunction to the formula `R v ~P` in node `v` of Fig. 3.4. However, given that `v` already contains `P`, it is clear that adding `not P` to `v` will end up in a contradiction (see Fig. 3.6). In this case it would be better to handle the disjunction by directly adding `R` to `v`.

This can be achieved by defining what is called *cut* rules. We have to account for four possible cases:

```
Rule CutOr_leftPos
  hasElement w or variable A variable B
  hasElement w not variable A

  add w variable B
  markExpressions w or variable A variable B Cut

Rule CutOr_leftNeg
  hasElement w or not variable A variable B
  hasElement w variable A

  add w variable B
  markExpressions w or not variable A variable B Cut

Rule CutOr_rightPos
  hasElement w or variable A variable B
  hasElement w not variable B

  add w variable A
  markExpressions w or variable A variable B Cut

Rule CutOr_rightNeg
  hasElement w or variable A not variable B
  hasElement w variable B

  add w variable A
  markExpressions w or variable A not variable B Cut
```

For example, the application of the rule CutOr_rightNeg to the premodel of Fig. 3.5 adds R to v.

Exercise 32 Define similar rules for negated conjunctions. In line with the preceding names these rules should be called CutNotAnd_left, CutNotAnd_right, CutNotAnd_left_not, and CutNotAnd_right_not.

Where should we put the CutOr_rightNeg rule in our strategy? Well, one should certainly try to apply it before the Or rule. So let us modify the strategy K_Strategy and insert all the above cut rules immediately before the Or rule. However, as you may check, the graphs are still duplicated by the Or rule. We can avoid this behaviour by replacing the Or rule by the Or_withRedundancyCheck rule of Sect. 3.10.1. Then our optimised boolean rules are as follows.

```
Strategy BooleanRules_Better
  Clash
  NotNot
  And
```

```
  NotOr
  CutOr_leftPos
  CutOr_leftNeg
  CutOr_rightPos
  CutOr_rightNeg
  CutNotAnd_left
  CutNotAnd_right
  CutNotAnd_left_not
  CutNotAnd_right_not
  Or_withRedundancyCheck
  NotAnd_withRedundancyCheck
end
```

Let us replace `BooleanRules` by `BooleanRules_Better` in `K_Strategy`. The resulting strategy avoids duplication of the graph of Fig. 3.4: LoTREC will only build a single graph for the input formula here.

However, consider the formula $(P \wedge R) \wedge (P \vee Q)$. With the above strategy we obtain `P & R` and `P v Q` in the first step (by the rule for conjunction). Observe that the redundancy check does not apply and that the premodel will therefore again be duplicated.

How can we delay the application of the `Or_withRedundancyCheck` rule as much as possible? Well, we should apply the non-duplicating rules as long as possible before any duplicating rule comes into play. This can be done in LoTREC by *nesting* two repeat-loops as follows.

```
Strategy BooleanRules_Optimised
  repeat
    Clash
    NotNot
    And
    NotOr
    CutOr_leftPos
    CutOr_rightPos
    CutOr_leftNeg
    CutOr_rightNeg
  end
  Or_withRedundancyCheck
  NotAnd_withRedundancyCheck
end
```

Let us call `K_Strategy_Optimised` the strategy resulting from the replacement of `BooleanRules` by `BooleanRules_Optimised`.

Exercise 33 Apply `K_Strategy_Optimised` to the formula

$$\neg\Box P \wedge \Box(P \vee Q) \wedge \Box(P \vee \neg Q)$$

Exercise 34 Apply `K_Strategy_Optimised` to the formula

$$(\neg P \vee \neg Q) \wedge (\neg P \vee Q) \wedge (P \vee \neg Q) \wedge (P \vee Q)$$

Explain the result by `LoTREC`'s simultaneous application of all applicable rules.

However not all of the optimizations we have introduced always allow us to avoid unnecessary premodel duplication. To see this consider the formula $P \vee (P \wedge Q)$ (which is equivalent to P): the rule `Or_withRedundancyCheck` applies to it and produces two premodels. The second is clearly redundant: if we can find a model for `P v (P & Q)` and `P & Q` then we can also find a model for `P v ( P & Q)` and `P`. However, none of our rules is able to block the application of the rule for disjunction.

Exercise 35 ([BdRV01], Sect. 6.7, p. 382) Consider the following two families of formulas:

$$B_i = (\Diamond p_i \wedge \Diamond\neg p_i)$$
$$S_i = \big((p_i \rightarrow \Box p_i) \rightarrow (\neg p_i \rightarrow \Box\neg p_i)\big)$$

where i is a natural number greater than or equal to zero. Let A_m be the conjunction of the following formulas:

- $B_1 \wedge \Box B_2 \wedge \cdots \wedge \Box^{m-1} B_m$;
- $\Box S_1 \wedge \cdots \wedge \Box^{m-1} S_1$;
- $\Box^2 S_2 \wedge \cdots \wedge \Box^{m-1} S_2$;
- $\ldots$;
- $\Box^{m-1} S_{m-1}$.

The expression $\Box^m$ stands for m repetitions of the modal operator $\Box$. For example:

$$\begin{aligned} A_3 = (\Diamond p_1 \wedge \Diamond\neg p_1) \wedge \Box(\Diamond p_2 \wedge \Diamond\neg p_2) \wedge \Box\Box(\Diamond p_3 \wedge \Diamond\neg p_3) \\ \wedge \Box\big((p_1 \rightarrow \Box p_1) \rightarrow (\neg p_1 \rightarrow \Box\neg p_1)\big) \\ \wedge \Box\Box\big((p_1 \rightarrow \Box p_1) \rightarrow (\neg p_1 \rightarrow \Box\neg p_1)\big) \\ \wedge \Box\Box\big((p_2 \rightarrow \Box p_2) \rightarrow (\neg p_2 \rightarrow \Box\neg p_2)\big) \end{aligned}$$

1. Call `K_Strategy` and `K_Strategy_Optimised` on A_1, A_2, A_3. Comment on the differences you observe.
2. What kind of graphs are built? What are the roles of p_i, B_i, S_i?
3. Prove that a model of A_m contains a tree containing at least 2^m nodes.
4. Compare the size of the formula A_m and the size of its model. What do you conclude?

3.11 Soundness, Termination, and... Completeness

Once we have entered a set of rules and a strategy combining these rules into the LoTREC interface we have achieved the definition of a *tableaux procedure*. The *input* of the procedure is a formula and the *output* is a set of saturated premodels: graphs which can no longer be modified by tableau rules. (Remember the definitions of rules and of closed, open and saturated premodels of Sect. 3.4.)

In the sequel we suppose that there is a given, fixed set of tableau rules. Let us formulate some desirable properties of strategies.

First of all, let us recall our initial motivation: given an input formula A, we want to build a Kripke model M such that $M, w \Vdash A$ for some possible world w in M. When every saturated premodel returned by the tableaux procedure is closed then we naturally expect that there is no Kripke model M wherein A is true in some world. The technical term for that is that the tableaux procedure should be *sound*.

Definition 10 (Soundness) A strategy is *sound* if and only if for every input formula A, if every saturated premodel for A is closed then A is unsatisfiable.

Second, termination is an important property of every strategy: if there are formulas for which the strategy loops then there is no hope to use that strategy in an automated way in order to solve reasoning problems e.g. such as program verification or the implementation of an artificial agent.

Definition 11 (Termination) A strategy is *terminating* if and only if for every input formula A, all premodels for A are saturated after a finite amount of time.

If ever our strategy is not terminating we might still use LoTREC in order to build models: we then have to use the interactive, step-by-step mode, as we already did in Sect. 3.6. This is similar to the way one proves theorems with assistants for higher-order logics such as Isabelle or HOL [Pau89, NPW02].

Third, when on input A a tableaux procedure outputs a graph M with root node w then we expect that M is indeed a Kripke model and that M, w $\Vdash$ A; or we at least expect that M can be transformed in a way such that such a model is obtained. The technical term for that is that the tableaux procedure should be *complete*.

Definition 12 (Completeness) A strategy is *complete* if and only if for every input formula A, the open saturated premodels that are built with the strategy can be transformed into Kripke models M such that $M, w \Vdash$ A for some w of M.

So Definition 12 does not require that the open saturated premodel that is returned by the tableaux procedure is directly a Kripke model: it only stipulates that the result can be transformed into a Kripke model. Such a transformation should be a systematic procedure.

The methods that we defined in this chapter are all sound, complete and terminating.

- We do not formally prove soundness of the tableau rules; it follows from the fact that our rules match the truth conditions. (The reader may consult Fitting's book [Fit83] or Goré's chapter in the Handbook of Tableaux Methods [Gor99] for detailed proofs.)
- In the next chapter we give a general termination criterion (Theorem 1) that covers all the methods that we have defined up to now.
- As to completeness, we refer the reader to the proofs in standard texts such as [Fit83, Gor99]. Instead, in Sect. 5.5 we are going to argue that from a practical point of view, soundness and termination of a model construction method are more important than completeness.

If a tableaux procedure is sound, complete, and terminating for a given logic then it provides a *decision procedure* for the model building problem, and therefore also for the satisfiability problem and for the validity problem. When it is possible to design a decision procedure for a logic then we say that the logic is *decidable*. We have already mentioned that first-order logic is undecidable. There also exist (rather complex) modal logics that are undecidable. We will not present such logics in this book. We refer the reader to more advanced textbooks on modal logic such as [CZ97, BdRV01, GKWZ03, BBW06] for more material on such logics.

3.12 Summary

In this chapter we have introduced the model construction method. It consists in decomposing the input formula by means of a set of rules mirroring the truth conditions in an order that is specified by the strategy. We have presented the tableaux procedure for the basic modal logic **K** as well as for its multimodal version $\mathbf{K}_n$ and the logic **ALC**. We have not only shown how rules and strategies work on paper, but also how they can be implemented in LoTREC.

The LoTREC platform differs from usual implementations of the tableaux method:

- "To apply a rule" means to simultaneously apply it to all nodes and all premodels, while this is not the case in the other implementations in the literature;
- Rule application is driven by a very simple strategy language: basically, the tableau rules are applied one after the other in a `repeat` loop, which terminates when none of the rules adds anything new; in contrast, tableaux implementations from the literature have a much more algorithmic flavour.

We ended the chapter with a discussion of three important requirements for a proof method: termination, soundness, and completeness.

Chapter 4
Logics with Simple Constraints on Models

We have seen in Sect. 1.9 that the models for several concepts naturally satisfy some constraints. Let us recall the card game of Sect. 1.3 where we modelled Ann's and Bob's knowledge by means of the graph of Fig. 1.8 that we repeat here as Fig. 4.1. In the 'Red' state the formula Red is true, and it is for that reason that the formula $\mathsf{K}_{\mathsf{Bob}}\neg\mathsf{Red}$ is false in that state. More generally, in appropriate models of knowledge the agents' accessibility relations should be reflexive. This allows us to exclude that A and $\mathsf{K}_I\neg A$ are true in the same state, whatever I and A are. This corresponds to the fact that $\mathsf{K}_I A \rightarrow A$ is valid in the class of reflexive models: knowledge is, by definition, true.

When we want to know whether a formula such as above is valid in the class of reflexive models then we are interested in the problem of validity in $\mathbf{C}$, where $\mathbf{C}$ is the class of reflexive Kripke models. (Remember that we have defined in Sect. 2.5 what the validity problem in $\mathbf{C}$ is.)

Unfortunately, the formula $\mathsf{Red} \wedge [\mathsf{Bob}]\neg\mathsf{Red}$ is satisfiable according to the tableaux procedure for $\mathbf{K}_n$ in Sect. 3.8: the procedure builds the model consisting of a single root node that is labelled by Red and having an empty accessibility relation for the agent Bob; then $[\mathsf{Bob}]\neg\mathsf{Red}$ is true at the root node because the root has no successors. The method of Chap. 3 does not account for reflexivity and is therefore not appropriate to tackle the model-building problem in the class of reflexive models.

So how can we take into account that models of knowledge have a reflexive accessibility relation? Well, one way to guarantee reflexivity is to add to each node a reflexive edge. Then the application of the rule for $\Box$ will close the premodel, as illustrated in Fig. 4.2.

Basically there are two different ways to deal with constraints on accessibility relations.

1. The constraint is enforced directly by explicitly updating the accessibility relation: the necessary reflexive, symmetric, transitive,... edges are added, just as we did in Fig. 4.2; this is typically the case in *labelled tableaux systems* [Gab02];

O. Gasquet et al., *Kripke's Worlds*, Studies in Universal Logic,
DOI 10.1007/978-3-7643-8504-0_4, © Springer Basel AG 2014

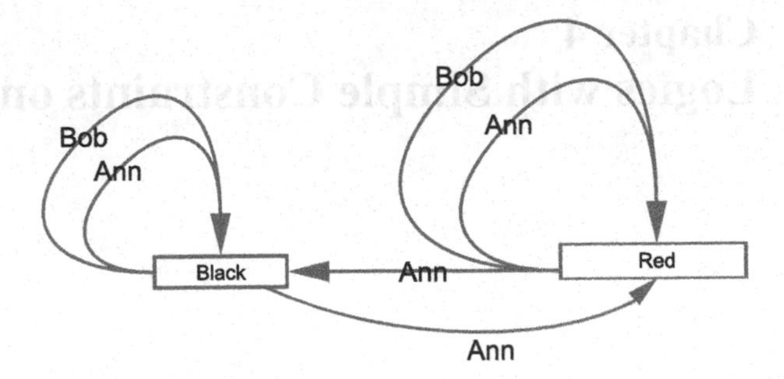

Fig. 4.1 Ann guessing the colour of Bob's card (actual world is the right Red world)

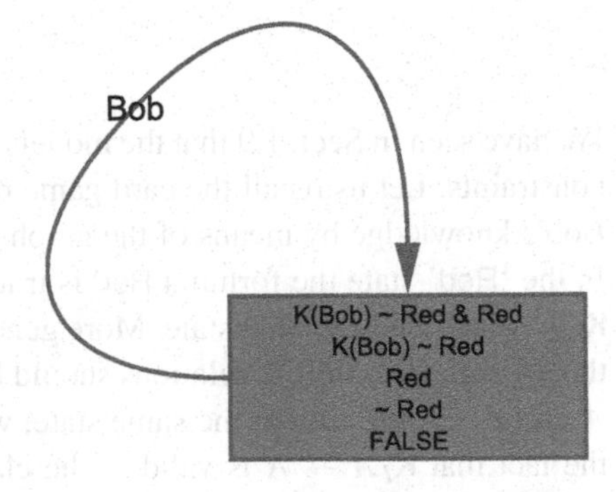

Fig. 4.2 The intended closed premodel for the formula $\mathsf{K}_{\mathsf{Bob}}\neg\mathsf{Red} \land \mathsf{Red}$

2. The constraint is enforced indirectly by modifying the way the modal connectives are analysed. This is the traditional way tableaux for modal logics work, as exemplified in Melvin Fitting's book [Fit83].

In both cases we only have to modify the rules dealing with the edges and/or with the modal connectives: the rules dealing with boolean connectives remain the same, and our strategies all call the strategy `BooleanRules` that we have defined in Sect. 3.6.4. (`BooleanRules` may be replaced by its optimised version `BooleanRules_Optimised` of Sect. 3.10.2.)

None of the modifications of the strategy `K_Strategy` in the present chapter endangers termination. This contrasts with the constraints that we shall investigate in Chap. 5, and for that reason we call the constraints of the present chapter *simple*.

The aim of this chapter is to adapt the tableaux method of the previous chapter in a way to capture several simple constraints on accessibility relations. We give the methods for logic $\mathbf{K}_2 \oplus$ **Inclusion** (Sect. 4.1), **KT** (Sect. 4.2), **KB** (Sect. 4.3), $\mathbf{K.Alt}_1$ (Sect. 4.4), **KD** (Sect. 4.5), **K.2** (Sect. 4.6) and finally **S5** (Sect. 4.7). In Sect. 4.8, we consider the basic hybrid logic **HL** which has a specific constraint on its valuation function. In the end of the chapter we give a general termination criterion which directly applies to all logics but **S5**, $\mathbf{S5}_n$, and **HL** (Sect. 4.9).

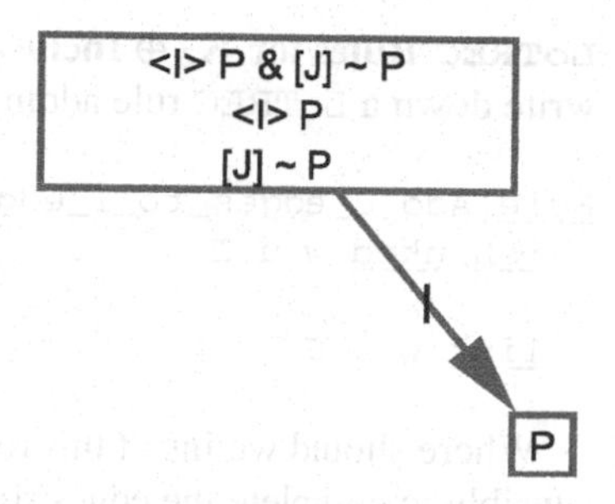

Fig. 4.3 An open premodel for $\langle I \rangle P \wedge [J]\neg P$ obtained with the method of Chap. 3

4.1 $\mathbf{K}_2 \oplus$ Inclusion: Inclusion of Relations

Consider a Kripke model where $R(I)$ and $R(J)$ are respectively the relations '*is a daughter of*' and '*is a child of*'. This is an example where we have *inclusion between two relations*. It is clear that if a property holds for all children then it holds also for daughters. We shall explain how to adapt the above model construction method in order to take this into account.

Suppose the set of indexes is $\mathcal{I} = \{I, J\}$, i.e., we have a multimodal language with two indexes I and J. Consider the logic of the class of frames such that $R(I) \subseteq R(J)$. In that class the axiom schema $\langle I \rangle A \rightarrow \langle J \rangle A$ is valid. This is an *interaction axiom* between two modal operators; we call it the **Inclusion** axiom. The $\langle I \rangle A \rightarrow \langle J \rangle A$ axiom schema can also be written using the dual modal operator as $[J]A \rightarrow [I]A$. The logic of that class of frames is called $\mathbf{K}_2 \oplus$ **Inclusion**.

As the formula $[J]A \rightarrow [I]A$ is a theorem of $\mathbf{K}_2 \oplus$ **Inclusion**, the formula $\langle I \rangle P \wedge [J]\neg P$ must be unsatisfiable. If we apply the strategy for the multimodal logic **K** of Sect. 3.8 to that formula then we obtain the open premodel of Fig. 4.3.

The two possible ways of modifying `K_Strategy` are the following.

1. We add a tableau rule which whenever there is an `I` edge between two nodes also adds a `J` edge;
2. We amend the tableau rule for the modal operator $[J]$ as follows: whenever some world `w` contains the formula `[J] A` then we not only propagate `A` to the `J`-successors of `w`, but also to all the `I`-successors of `w`.

The first option *explicitly* ensures that `I` edges are also `J` edges, while the second option does so only *implicitly*. The latter can be viewed as a modification of the truth condition for $[J]$ to:

$$M, w \Vdash [J]A \quad \text{iff for every world } u, \text{ if } \langle w, u \rangle \in R(J) \text{ or } \langle w, u \rangle \in R(I)$$
$$\text{then } M, u \Vdash A$$

Note that graphs that are built in this way do not satisfy the inclusion constraint. However, the required `J` edges may be added later without harm: our modified tableau rule for $\Box J$ is designed in a way such that the same formulas as before will be true in the resulting model.

LoTREC Rules for $\mathbf{K}_2 \oplus$ Inclusion—Explicit Inclusion It is straightforward to write down a LoTREC rule adding J edges whenever there are I edges:

```
Rule Add_J_edges_to_I_edges
  isLinked w u I

  link w u J
```

Where should we insert this rule into our strategy for multimodal $\mathbf{K}_n$? Well, it is sensible to complete the edges right after they are created by the Pos and NotNec rules and before they are exploited by the Nec rule:

```
Strategy K2_Inclusion_Strategy
  repeat
    BooleanRules
    Pos
    NotNec
    Add_J_edges_to_I_edges
    Nec
    NotPos
  end
end
```

Exercise 36 Run the above strategy on the unsatisfiable formula $\langle I \rangle P \wedge [J]\neg P$. Check that LoTREC finds a clash.

LoTREC Rules for $\mathbf{K}_2 \oplus$ Inclusion—Implicit Inclusion The following tableau rule for the modal operator $[J]$ takes the inclusion of $R(J)$ in $R(I)$ into account in an implicit way:

```
Rule Nec_J_to_I_successors
  hasElement w nec J variable A
  isLinked w u I

  add u variable A
```

Let us insert this rule somewhere in the strategy defined in Sect. 3.7. Any position will do, for example right after the Nec rule. If we run the resulting strategy on $\langle I \rangle P \wedge [J]\neg P$ then we obtain the closed premodel of Fig. 4.4.

Exercise 37 Consider the formula $A = \langle I \rangle P \wedge \neg\langle J \rangle P$.

1. Check by hand that A is unsatisfiable in models where $R(I) \subseteq R(J)$.
2. Use LoTREC and try to build a model for A with the above rules and strategy. Why does LoTREC find a model?
3. Add the missing rule.
   (Hint: it should be called NotPos_J_To_I_Successors.)

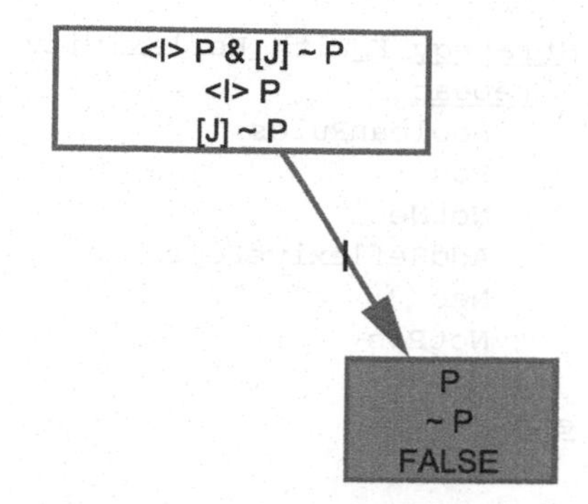

Fig. 4.4 A closed premodel for $\langle I\rangle P \wedge [J]\neg P$ that is obtained by application of the rule `Nec_J_to_I_¬successors`

Exercise 38 Let $\mathcal{I} = \{I, J, K\}$. Consider the logic of the class of models such that $R(I) \subseteq R(J) \subseteq R(K)$. Its axiomatisation can be obtained by adding to the axiomatisation of $\mathbf{K}_3$ the axioms **Inclusion**(I, J) and **Inclusion**(J, K).

Design (implicit or explicit) tableau rules for that logic and a define a strategy combining them.

4.2 KT: Reflexivity

KT is the monomodal logic of reflexive models. It is characterized by the axiom **T**: $\Box A \to A$ (see Sect. 2.6).

As before, given that we only have one modal operator we write $\Box$ and $\Diamond$ instead of $[I]$ and $\langle I\rangle$ for the modal operators and R instead of $R(I)$ for the accessibility relation.

We have shown in the introduction of this chapter that the premodels that are returned by the tableaux procedure of Chap. 3 are irreflexive. Again, we can either add reflexive edges explicitly or simulate them by a supplementary rule for $\Box$.

LoTREC Rules for KT—Explicit Reflexivity The following rule adds to each node the required reflexive:

```
Rule AddReflexiveEdges
  isNewNode w

  link w w R
```

The condition `isNewNode w` is required because the condition part of a rule cannot be empty (else the rule would not be a legal LoTREC rule, cf. Definition 9 in Sect. 3.5, page 68). As its name says, `isNewNode w` checks that the node w has either just been created or has just been modified by the addition of a formula or an edge connecting it to another node. So a rule having `isNewNode w` in its condition part is applicable—only once—to each node w that is added to the graph.

The rule `AddReflexiveEdges` can be called anywhere inside the outermost `repeat` loop of the strategy of Sect. 3.7; for example:

```
Strategy K_With_Reflexivity
  repeat
    BooleanRules
    Pos
    NotNec
    AddReflexiveEdges
    Nec
    NotPos
  end
end
```

LoTREC Rules for KT—Implicit Reflexivity Instead of explicitly adding reflexive edges the reflexivity constraint can be simulated by changing the truth condition of the $\Box$-connective as follows:

$$M, w \Vdash \Box A \quad \text{iff } M, w \Vdash A \text{ and for all } u \text{ such that } \langle w, u\rangle \in R, M, u \Vdash A$$

So a $\Box$-formula is true in a given world w of a model if and only if it is true both at every R successor of w and in the world w itself. We can implement this by the following LoTREC rule:

```
Rule Nec_forReflexivity
  hasElement w nec variable A

  add w variable A
```

Exercise 39 Implement the dual rule `NotPos_forReflexivity` in LoTREC.

The rules `Nec_forReflexivity` and `NotPos_forReflexivity` have to be inserted into `K_Strategy`, for example right after the rules for the boolean connectives. Call the result `KT_Strategy`. The premodels that are returned by the implicit strategy are irreflexive. If ever we want to obtain a reflexive model we therefore still have to insert the above rule `AddReflexiveEdges` at the end of `KT_Strategy`, after the `repeat` loop.

Exercise 40 Run the `KT_Strategy` on the formulas $P \wedge \Box\neg P$ and $P \wedge \Diamond\neg P$.

Exercise 41 Consider the bimodal logic $\mathbf{K}\otimes\mathbf{KT}$, i.e. the logic with two modalities: a **K**-modality, say $[I]$, and a **KT**-modality, say $[J]$. So the class of models is such that $R(I)$ is arbitrary and $R(J)$ is reflexive.

1. Find a set of tableau rules and a strategy for this logic.
2. Check the formula $A_0 = \neg P \wedge [I](P \vee Q) \wedge [J]\neg Q$.
3. Adapt the rules and the strategy to deal with the inclusion axiom: $[J]A \rightarrow [I]A$ of Sect. 4.1. Is A_0 still satisfiable?

4. Adapt the rules of Item 1. to take into account the inclusion constraint $R(J) \subseteq R(I)$, which corresponds to the inclusion axiom $[I]A \to [J]A$. What about the satisfiability of A_0 in this logic?

Exercise 42 Consider the class of frames satisfying the weak reflexivity constraint: $\langle w, u \rangle \in R$ implies $\langle u, u \rangle \in R$, for every w, u.

1. Which axiom schema is valid in these models?
2. Argue why it is a reasonable principle of a logic of belief.
3. Find the LoTREC rule for that logic.

Exercise 43 Consider the bimodal logic $\mathbf{K}_2 \otimes \mathbf{KT}$, i.e. the logic with three modalities: two **K**-modalities, say $[I]$ and $[J]$, and a **KT**-modality, say $[K]$. The class of models is such that $R(I)$ and $R(J)$ are arbitrary and $R(K)$ is reflexive.
Define a tableaux procedure for that logic.

Exercise 44 Give the model construction method for a multimodal logic $\mathbf{K}_n \otimes \mathbf{KT}$, i.e. a logic with n **K**-modalities and one **KT**-modality.

4.3 KB: Symmetry

Many relations are intuitively symmetric, such as 'being married to' or 'being a sibling of'. The neighbourhood relation between countries is another example of a symmetric relation: graph nodes are countries, and there is an edge from country w to country u if and only if w is a neighbour of u. Whenever there is an edge from w to u then there is also an edge from u to w. Suppose that when the atomic formula EU labels node w then this means that w is a member state of the European Union. Hence when the formula $\neg\mathsf{EU} \land \Diamond\Box\mathsf{EU}$ is true at node w then country w is not a member of the European Union and has a neighbour, say u, such that all of u's neighbours are EU members. Our procedure should allow us to find out that this cannot be the case. This contrasts with modal logic **K**: as the reader may check by applying the tableaux method of Chap. 3, the formula $\neg\mathsf{EU} \land \Diamond\Box\mathsf{EU}$ has a Kripke model (that is non-symmetric).

In this section, we explain how to tune the basic model construction procedure of Chap. 3 to obtain a model with a symmetric accessibility relation. As before, for simplicity we let the set of edge labels $\mathcal{I}$ be a singleton, i.e., there is a single accessibility relation R. That relation is *symmetric* if the constraint $R \subseteq R^{-1}$ holds (see Sect. 2.6).

LoTREC Rules for KB—Explicit Symmetry In order to make the above formula unsatisfiable we add the following rule.

```
Rule AddSymmetricEdges
  isLinked w u R
```

```
link u w R
```

We may insert this rule anywhere into the strategy; say just after `Pos` and before calling `Nec`.

Remark 10 One may think that we need to add the condition `isNotLinked u w R` (which exists in LoTREC) to make sure that such a link does not exist. This is however built-in in LoTREC: two nodes cannot be linked by two edges of the same label.

LoTREC Rules for KB—Implicit Symmetry An implicit way of taking the symmetry constraint into account is by propagating the $\Box$-formulas from children nodes to their parent nodes, as if they were linked to them. This corresponds to the following truth condition:

$$M, w \Vdash \Box A \quad \text{iff } M, u \Vdash A \text{ for all } u \text{ such that } \langle w, u \rangle \in R \text{ or } \langle u, w \rangle \in R$$

The following rule does the job:

```
Rule Nec_forSymmetry
  hasElement w nec variable A
  isLinked u w R

  add u variable A
```

This rule has to be added to `K_Strategy`, e.g. right after the `Nec` rule.

Just as for the implicit version of the tableaux procedure for **KT**, if we want to transform an open saturated premodel into a model we have to insert the rule `AddSymmetricEdges` after the `repeat` loop.

Exercise 45 Let $\mathcal{I} = \{I, J\}$. Consider the class of models such that $R(J) = R(I)^{-1}$. This is the so-called tense extension of the basic modal logic **K**. A common notation for that logic is the so-called tense extension of **K**, noted $\mathbf{K}_t$. $\mathbf{K}_t$ is for example the logic of the temporal connectives "tomorrow" and "yesterday."

Find the LoTREC rules for that logic.

4.4 K.Alt$_1$: Partial Function

Let $\mathcal{I}$ be a singleton yet again. In this section, we consider the class of frames where the relation R is a partial function. This means that a given world can have at most one R-successor.

In general, a world in a Kripke frame may have several successors. This can be illustrated syntactically by the fact that the formula $\Diamond P \wedge \Diamond \neg P$ is satisfiable in

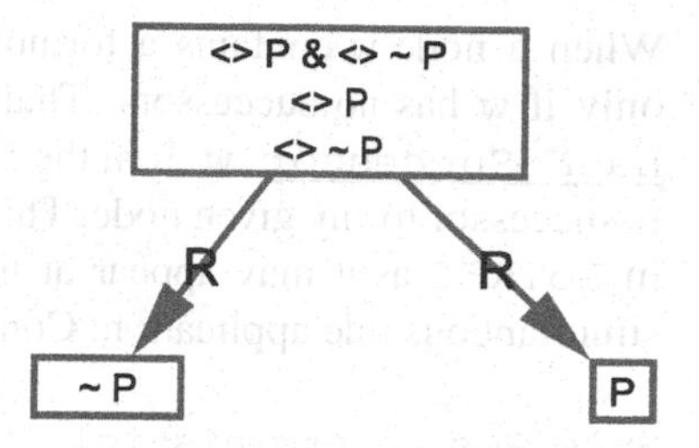

Fig. 4.5 Open **K** premodel for the formula $\Diamond P \wedge \Diamond \neg P$ obtained with `K_Strategy` of **K** (Chap. 3)

modal logic **K**. Figure 4.5 contains the open saturated premodel that is obtained by `K_Strategy`. On the contrary, in the class of deterministic Kripke frames a world can have at most one successor. The logic of that class is usually called **K.Alt**$_1$, where **Alt**$_1$ stands for "there is at most one alternative (to the actual world)." The formula $\Diamond P \wedge \Diamond \neg P$ is no longer satisfiable in **K.Alt**$_1$.

Relations with at most one successor can be used to model deterministic programs: while nondeterministic programs can be viewed as relations (on program states), deterministic programs are functions. These functions are partial because a program may not be executable in some states. If we suppose that time is linear then deterministic relations can also be used to model the temporal 'next' relation. In databases, some relations have to be partial functions, for example '*husband of*' or '*social security number of*.'

How can we produce a clash for the formula $\Diamond P \wedge \Diamond \neg P$? First of all, when a node `w` contains `<> P & <> ~P` then it will contain both `<> P` and `<> ~P` by the tableau rule for conjunction. We then should either create a unique successor of `w` containing both `P` and `~P` as shown in Fig. 4.6, or we should ensure that all `R`-successors have the same node labels. We start with the latter recipe because it is simpler.

LoTREC Rules for K.Alt$_1$—Implicit Functionality The rule is simple: whenever we encounter a `pos A` then we not only create a new successor and add `A` to it—which is what the `Pos` rule already does—, but also add `A` to *all* successors.

```
Rule Pos_Everywhere
  hasElement w pos variable A
  isLinked w u R

  add u variable A
```

Together with the dual rule `NotNec_Everywhere` this rule has to be added to `K_Strategy`, e.g. right after the `Pos` rule.

This is however a rather non-minimalistic way of ensuring the functionality constraint, and it might look a bit strange to the reader: the `Pos_Everywhere` rule leads to multiple successors that are all labelled by the same formulas.

LoTREC Rules for K.Alt$_1$—Explicit Functionality The first thing to do is to drop the rule `Pos` for the $\Diamond$ operators of Chap. 3: they generate multiple successors.

When a node w contains a formula <> A then we should add a successor to w only if w has no successors. That is why we should use an additional condition hasNoSuccessor w R in the Pos rule in order to avoid creating more than one R-successor to any given node. This is however not as straightforward to implement in LoTREC as it may appear at first glance. The reason is LoTREC's politics of simultaneous rule application. Consider the following two rules for the ◇ operator.

```
Rule Pos_deterministic1
  hasElement w pos variable A
  hasNoSuccessor w R

  createNewNode u
  link w u R

Rule Pos_deterministic2
  hasElement w pos variable A
  isLinked w u R

  add u variable A
```

The first rule says: if w has no successors and contains some <> A then create one. The second rule says: if there is a <> A at w having already a successor u then add A to that world u. Note that the first rule need not add A since this is what the second rule does.

Let us replace the Pos rule of K_Strategy by the two above rules, plus the dual rules NotNec_deterministic1 and NotNec_deterministic2. It turns out that this does not work as the reader might have expected: it again delivers the premodel of Fig. 4.5. The cause is that LoTREC applies a given rule to all matching patterns at once (cf. Sect. 3.6.1), and when we have two ◇-formulas in a given node without successors then *each of them* will trigger the application of Pos_deterministic1 because the node has no successors at the time of the test! So the rule will be applied twice, once per ◇-formula, and will yield two successors just as the old Pos rule.[1]

A way out is to insert the strategy keyword applyOnce just before the name of the rule Pos_deterministic1. This allows the application of the rule on only one matching pattern at a time. Encapsulating the applyOnce Pos_deterministic1 call inside a repeat ... end routine applies the rule on all possible occurrences, but after considering them one-by-one, which means that the condition hasNoSuccessor now does the job we expect. Therefore, when the rule is applied to one of the ◇-formulas it creates a successor; it then

[1] So the hasNoSuccessor condition is only effective in subsequent applications of the Pos_deterministic1 rule: when some other ◇-formulas are added to the same node then hasNoSuccessor will make the Pos_deterministic1 rule fail. So it would for example succeed in closing the premodel for <> P & (<> ~P & Q). (Observe that parentheses are relevant: (<> P & <> ~P) & Q would not close.)

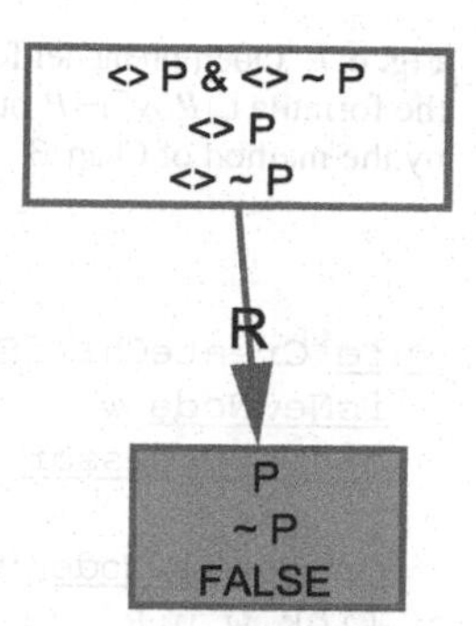

Fig. 4.6 The intended closed premodel for $\Diamond P \land \Diamond\neg P$ in **K.Alt**$_1$

becomes inapplicable again (due to the presence of that successor) for all the other $\Diamond$-formulas: only one $\Diamond$-formula will be processed.

This said, the strategy is as follows:

```
Strategy Kalt1_Strategy
  repeat
    BooleanRules
    applyOnce Pos_deterministic1
    applyOnce NotNec_deterministic1
    Pos_deterministic2
    NotNec_deterministic2
    Nec
    NotPos
  end
end
```

Running this strategy on the formula $\Diamond P \land \Diamond\neg P$ delivers the premodel of Fig. 4.6, containing a clash as expected.

4.5 KD: Seriality

A relation is serial if each world has at least one successor. One of the applications of a serial relation is the representation of obligations, as we have shown in Sect. 1.4. The most prominent logic of obligation is the so-called *standard deontic logic* (**SDL**). It is characterised by the **D** axiom $\Box A \rightarrow \Diamond A$—where **D** stands for 'deontic'—which says that what is obligatory is also permitted. This is sometimes also called the idealisation axiom.

The negation of the **D** axiom $\neg(\Box A \rightarrow \Diamond A)$ is equivalent to $\Box A \land \Box\neg A$. We have seen in Sect. 3.7 that `K_Strategy` returns the open saturated premodel of Fig. 4.7.

To enforce seriality one should make sure that every possible world can access at least one other world. In a first try this could be ensured by adding a rule to those for **K** that adds a successor to any node without `R` successors, as follows:

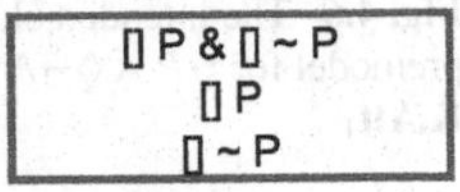

Fig. 4.7 Open premodel for the formula $\Box P \wedge \Box\neg P$ built by the method of Chap. 3


```
Rule CreateChildForLeafs_tentative
  isNewNode w
  hasNoSuccessor w R

  createNewNode u
  link w u R
```

The condition hasNoSuccessor w R verifies that the node w has no successor by the relation R. The node w has to be instantiated by other conditions. Here, w is instantiated by the condition isNewNode w.

This is however a dangerous rule: it keeps on creating new nodes, and therefore makes LoTREC loop. (You have to stop LoTREC 'by hand' by closing the window.)

LoTREC Rules for KD Let us further restrict the condition part of our above rule in order to avoid loops: we require that the node must contain at least one nec formula.

```
Rule CreateChildForLeafsWith_Nec
  hasElement w nec variable A
  hasNoSuccessor w R

  createNewNode u
  link w u R
```

The condition hasElement w nec variable A succeeds whenever w contains *any* nec-formula. The condition hasNoSuccessor w R succeeds only if w has no R successors.

Let us insert this rule into K_Strategy of Sect. 3.7, together with the dual rule CreateChildForLeafsWith_NotPos. When we run the resulting strategy then we can see that the rule CreateChildForLeafsWith_Nec creates as many successors for a given leaf node as there are $\Box$-formulas in it. This comes with LoTREC's simple definition of rule application: it is the price to pay for LoTREC's simplicity.

Note that there is actually nothing *wrong* with this: if we continue to run the strategy on the premodel of Fig. 4.8 we will obtain a closed premodel. It is however a bit annoying to have a strategy that leads to models that are bigger than needed. To avoid this we may prefix the rule by the keyword applyOnce. The strategy then is as follows:

```
Strategy KD_Strategy
  repeat
```

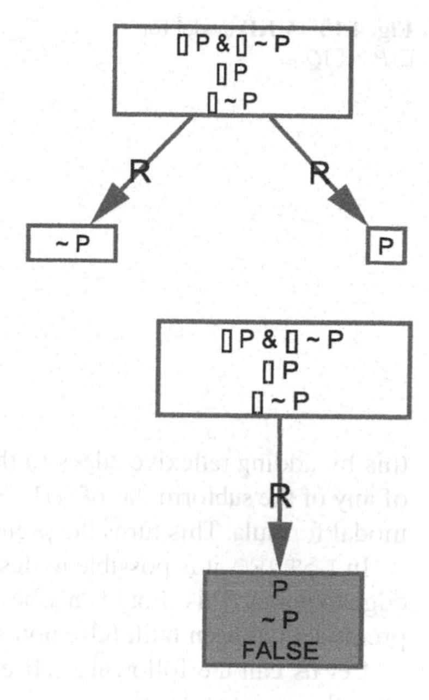

Fig. 4.8 The premodel after the call of the `CreateChildForLeafsWith_Nec` rule. Two nodes are created (one per □-formula)

Fig. 4.9 A closed premodel for $\Box P \land \neg \Diamond P$

```
    BooleanRules
    Pos
    NotNec
    Nec
    NotPos
    applyOnce CreateChildForLeafsWith_Nec
    applyOnce CreateChildForLeafsWith_NotPos
  end
end
```

The result of applying `KD_Strategy` on our example formula is depicted in Fig. 4.9.

Exercise 46 Run `KD_Strategy` on the formula $\Diamond P \land \Diamond \neg P$.

Remark 11 Another—indirect—way to ensure seriality for the relevant nodes is to add a rule adding $\Diamond \top$ to every node; or rather, to every node containing some $\Box A$ or $\neg \Diamond A$ because the strategy would not terminate otherwise. This would avoid the complications of the above rule. However, such a rule is not what is called analytic: it adds formulas that are not subformulas of some formula existing in the graph. Therefore our general termination theorem—Theorem 1 of Sect. 4.9—would not apply.

Turning an Open Saturated Premodel into a KD Model In the last exercise we built an open saturated premodel for $\Diamond P \land \Diamond \neg P$. This is however not a legal **KD** model yet because the leaves of our graph have no successors. It is easy to remedy

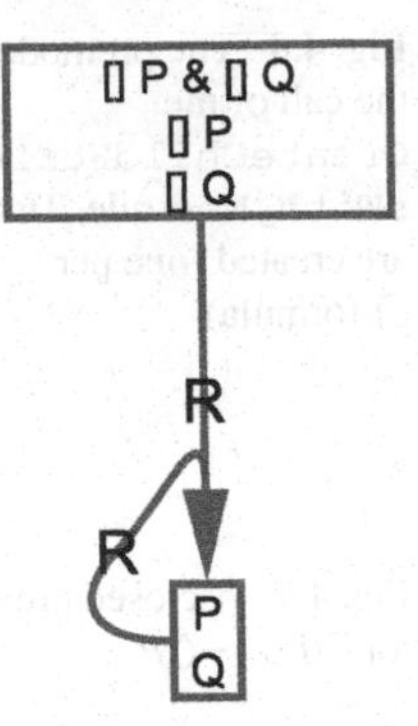

Fig. 4.10 A **KD** model for $\Box P \wedge \Box Q$

this by adding reflexive edges to these leaves. This will not change the truth value of any of the subformulas of $\Diamond P \wedge \Diamond \neg P$ because the leave nodes do not contain any modal formula. This turns the premodel into a truly serial model.

In LoTREC, it is possible to design a rule that does this job of adding a reflexive edge to nodes. This should only be done *after* the repeat loop, once an open saturated premodel has been built (else non-termination would be back again).

Let us call the following self-explanatory rule at the end of our strategy, right after the `repeat` loop:

```
Rule AddReflexiveEdgesToLeafNodes
  isNewNode w
  hasNoSuccessor w R

  link w w R
```

The conditions `isNewNode w` and `hasNoSuccessor w R` ensure that the rule is applied on new nodes which have no R successors. The overall strategy is therefore the following.

```
Strategy KD_Strategy_Plus_Model
  KD_Strategy
  AddReflexiveEdgesToLeafNodes
end
```

If we run `KD_Strategy_Plus_Model` on the formula $\Box P \wedge \Box Q$ then we obtain the premodel of Fig. 4.10. In the same way as in Sect. 3.2.3, we can transform the open premodel into a model, viz. by simply dropping the complex node labels. The resulting model is serial and the initial formula $\Box P \wedge \Box Q$ holds in its initial world.

Exercise 47 Combine the tableaux procedure for $\mathbf{K.Alt}_1$ and that for **KD** and write a tableaux procedure for the logic where the relation is a *total function*, i.e., where every node has *exactly one* successor. Try to find a strategy avoiding the keyword `applyOnce`.

Exercise 48 Write a tableaux procedure for the logic of the class of binary trees, i.e., the class of Kripke frames where a node has exactly two successors.

4.6 K.2: Confluence

An accessibility relation R is said to be confluent if and only if $\langle w, u\rangle, \langle w, v\rangle \in R$ implies that there exists a world x such that $\langle u, x\rangle \in R$ and $\langle v, x\rangle \in R$. We call the world x the confluence world of the tuple (w, u, v). The class of confluent frames can be axiomatised by adding the following axiom schema to the axiomatisation of **KD**:

$$.2 : \Diamond\Box P \to \Box\Diamond P$$

The axiom is sometimes called **G** in honour of Peter Geach. It has been studied as a principle of the logic of knowledge [Len78, Len95]. In the present section we consider the basic logic of confluent frames: the modal logic **K.2**.

A First Non-terminating Attempt A straightforward way to deal with confluence to add the following rule to those for modal logic **K**:

```
Rule Confluence_naive
  isLinked w u R
  isLinked w v R

  createNewNode x
  link u x R
  link v x R
```

The rule completes the premodels with the required confluence worlds.

Let us insert the confluence rule into the strategy `K_Strategy` of Sect. 3.7 and let us run it with the formula $\Diamond P \wedge \Diamond Q$ in the step-by-step mode. Part of LoTREC's output is shown in Fig. 4.11. As the reader may observe, the `Confluence_naive` rule is applicable on the four empty leaf nodes of step 2. This triggers the creation of 18 empty leaves of step 3. This should suffice to convince the reader that the strategy loops, just as the addition of the naive rule for seriality did.

LoTREC Rules for K.2 Just as in the case of seriality, termination can be enforced by requiring that `u` or `v` are non-empty. (We may actually add both conditions and require that both are non-empty: it will not vitiate completeness.) The rule then becomes:

```
Rule Confluence
  isLinked w u R
  isLinked w v R
  hasElement u variable A
```

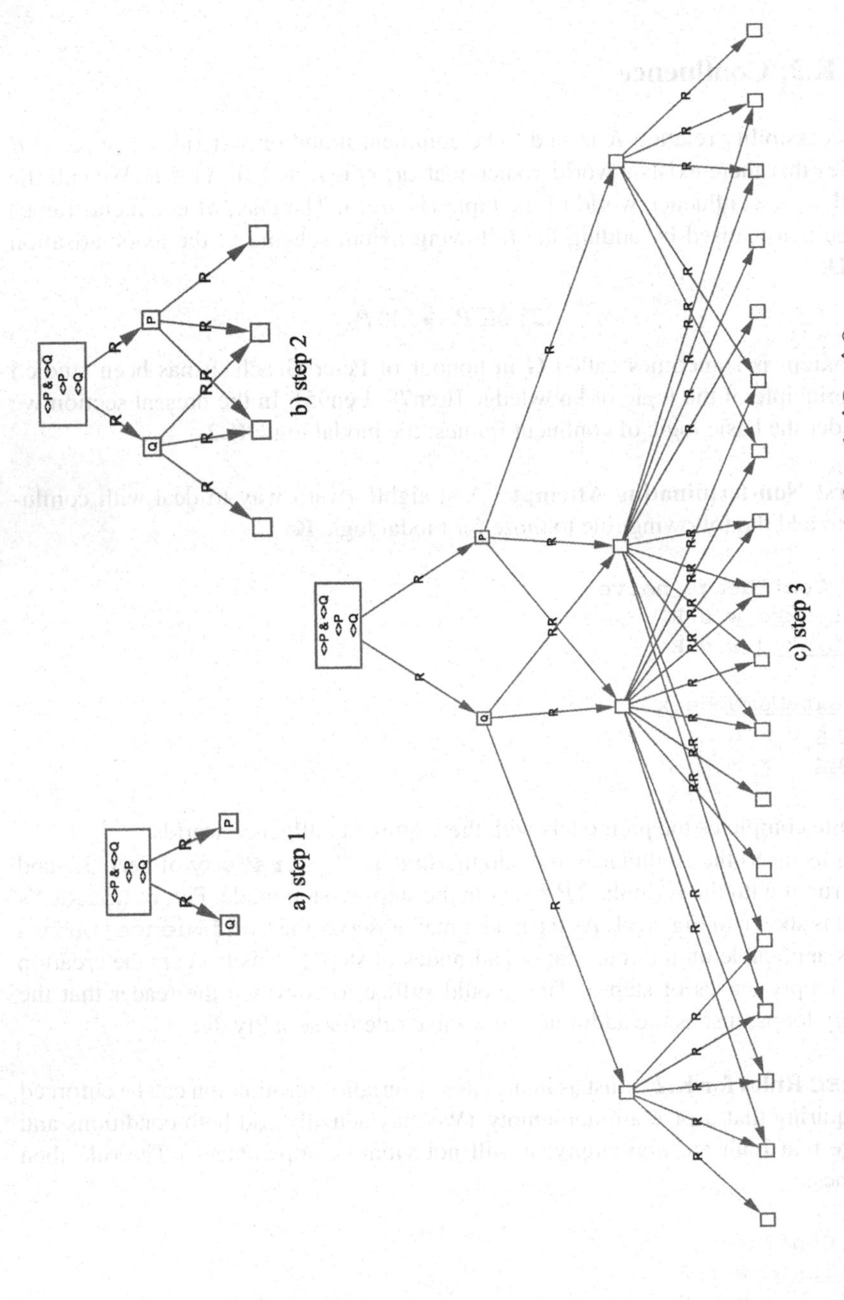

Fig. 4.11 The first three steps of running a strategy with rule `Confluence_naive` on the formula $\Diamond P \land \Diamond Q$

```
hasElement v variable B

createNewNode x
link u x R
link v x R
```

Let us add `Confluence` to `K_Strategy` and run it on $\Diamond P \wedge \Diamond Q$. The construction terminates at step 2 of Fig. 4.11 without entering step 3: the termination problem is solved.

The procedure can be shown to be complete. However, step 2 of Fig. 4.11 has four confluence nodes, while addition of a single confluence node would have been enough to make the premodel of step 1 confluent. From now on until the end of the section, we try to optimise the tableaux procedure for **K.2** in order to reduce the number of confluence nodes. Any reader who is happy with the terminating and complete procedure as it stands may safely skip the rest of this section.

Optimising the `Confluence` Rule Let us try to understand why there are so many nodes created by the `Confluence` rule. To that end we temporarily add two further actions to it:

```
Rule Confluence_debug
  isLinked w u R
  isLinked w v R
  hasElement u variable A
  hasElement v variable B

  createNewNode x
  link u x R
  link v x R
  add x nodeVariable u
  add x nodeVariable v
```

The LoTREC expression `nodeVariable u` designates node u's internal name: the name LoTREC assigns to the node variable u during the computation. The action `add x nodeVariable u` adds that internal name to the confluence node x. This allows one to memorise where x comes from. Let us run a strategy with this rule—again in a step-by-step mode—on the formula of Fig. 4.11: we obtain the premodel of Fig. 4.12 at the second step.

In this premodel, the user may check (by right-clicking on the nodes) that `node2` is the node on the right of the second row, containing `P`, and that `node3` is the node on the left of the second row, containing `Q`. On the third row, there are four nodes that have been created by the rule of confluence `Rule Confluence_debug`, from the left to the right:

- a node labelled `node3`, where both `u` and `v` were the same node `node3`;
- a node labelled `node3` and `node2`, where `u` was the node `node3` and `v` was the node `node2`;

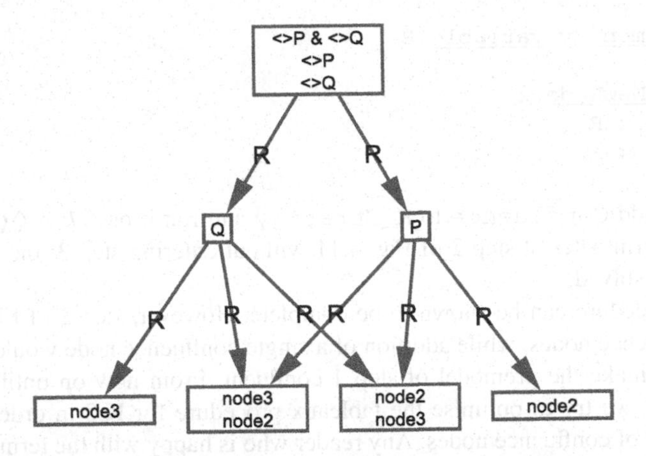

Fig. 4.12 Debugging the confluence rule: tracing the origins of the confluence nodes

- a node labelled `node2` and `node3`, where `u` was the node `node2` and `v` was the node `node3`;
- a node labelled `node2`, where both `u` and `v` were the same node `node2`.

Two of these nodes were created because (`node2`, `node3`) and (`node3`, `node2`) are considered to be different possible instances for the couple of variables (`u`, `v`). The creation of only one of these two nodes would however suffice to ensure confluence. So how could we ensure that we create only one confluent node per couple of nodes with a common parent node?

What we should do is to write rules in a way such that the patterns (`node2`, `node3`) and (`node3`, `node2`) are considered to be the equivalent. We can do the following in order to prevent the application of a rule to a pattern whose equivalent pattern (modulo a given criterion) was already matched:

1. Annotate the nodes by adding actions that highlight information indicating which patterns were taken into account (by means of additional actions);
2. Prevent the confluence rule from being applied on patterns which have already been taken into account (by means of additional—negative—conditions);
3. Guarantee that the rule is considering the patterns not in parallel but one by one (by means of the <u>`applyOnce`</u> keyword).

In our example, when a couple of variables (`u`, `v`) is instantiated by a couple of instance nodes (`Instance1`, `Instance2`), then the confluence rule should not consider the equivalent instantiation by (`Instance2`, `Instance1`). In order to do so we introduce a new auxiliary connective `done` which we declare to have two arguments and to be displayed as `Done(_,_)`. It allows us to refine the confluence rule: we are going to add to each confluence node information about the couples of nodes that were responsible for its creation (and that were therefore already taken into account). Here is the modified rule:

```
Rule Confluence_Optimised
```

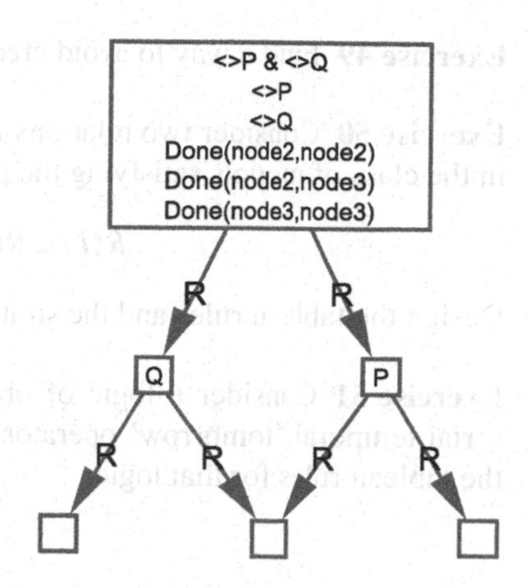

Fig. 4.13 Running `Confluence_Strategy` on $\Diamond P \wedge \Diamond Q$

```
isLinked w u R
isLinked w v R
hasElement u variable A
hasElement v variable B
hasNotElement w done nodeVariable u nodeVariable v
hasNotElement w done nodeVariable v nodeVariable u

createNewNode x
link u x R
link v x R
add w done nodeVariable u nodeVariable v
```

The condition

```
hasNotElement w done nodeVariable u nodeVariable v
```

checks that `w` does not contain the expression `done nodeVariable u nodeVariable v`. Observe that all the variables of this condition should be instantiated by other conditions in order to make the rule a legal LoTREC rule (cf. Definition 9 in Sect. 3.5, page 68). In the strategy, this rule should be prefixed by the `applyOnce` keyword:

```
Strategy Confluence_Strategy
  repeat
    BooleanRules
    Pos
    Nec
    applyOnce Confluence
  end
end
```

That strategy returns the premodel of Fig. 4.13 when called on $\Diamond P \wedge \Diamond Q$.

Exercise 49 Find a way to avoid creating the empty leaf nodes of Fig. 4.13.

Exercise 50 Consider two relations $R(I)$ and $R(J)$, and suppose we are interested in the class of models satisfying the permutation constraint

$$R(I) \circ R(J) \subseteq R(J) \circ R(I)$$

Design the tableau rules and the strategy for that logic.

Exercise 51 Consider a logic of obligation and time having a deterministic and serial temporal 'tomorrow' operator X such that $\mathsf{O}_I\mathsf{X}A \to \mathsf{X}\mathsf{O}_I A$ is valid. Design the tableau rules for that logic.

4.7 S5: A Single Equivalence Relation

In Sect. 1.3 we saw that an agent's knowledge can be modelled by an equivalence relation. Such relations are usually defined by the constraints of reflexivity, transitivity and symmetry, but they can also be defined by reflexivity, transitivity and euclideanity. Therefore **KT4B** and **KT45** are the same logic; however, the logic of a single equivalence relation is traditionally rather called **S5**.

In this section we develop a tableaux procedure for monomodal **S5**. Its set of labels is a singleton: $\mathcal{I} = \{I\}$. We present it therefore in terms of the standard modal operator K of epistemic logic instead of $\Box$. In order to simplify the presentation we suppose that there is no dual operator for K in the language. We avoid in this way writing down tableau rules for the case of the dual possibility operator $\hat{\mathsf{K}}$. According to the conventions of Table 2.3 (page 37), K is written `knows` and is displayed as `K` in LoTREC.

LoTREC Rules for S5—Implicit Equivalence Classes The constraints on R can be accounted for both in an explicit and in an implicit way. The explicit solution seems to be easier and more straightforward than the implicit solution: adding reflexive and symmetric edges is simple, as we have already seen in Sects. 4.2 and 4.3. Adding the transitive edges can also be implemented by a simple tableau rule. However, the latter is dangerous: the model building process may not terminate, as we shall show in Sect. 5.1. by means of the example formula $\mathsf{K}\neg\mathsf{K}P$.

We therefore only present the implicit solution here. It consists in representing an equivalence class by means of a tree of depth one: the root is linked to its children. We moreover add reflexive edges to all children. This allows one to identify the root as the only node without a reflexive edge. (The other way round, we could as well identify the root node as the only node having a reflexive edge; this is however less straightforward to generalise to the multimodal version of **S5**.) This trick allows us to give a special role to the root: when some of its children is labelled by $\neg\mathsf{K}A$ then we are not going to create a new successor of that child, but we are rather going to create a new successor of the root. This allows us to better control the graph construction.

Formulas of the form ¬K*A* are handled by two rules. The first sends all formulas of the form `not knows A` back to the root, and the second creates new child nodes.

```
Rule NotNec_copyToRoot
  hasElement w not knows variable A
  isLinked w0 w R

  add w0 not knows variable A

Rule NotNec_fromRoot
  hasElement w0 not knows variable A
  isNotLinked w0 w0 R

  createNewNode u
  link w0 u R
  link u u R
  add u not variable A
```

Formulas of the form K*A* should be propagated to every node in the equivalence class. We have three rules for that: a rule sending `knows A` formulas to the root, another rule which is nothing but the standard `Nec` rule for □ (replacing □ by the epistemic operator `knows`), and a rule accounting for the reflexivity of the accessibility relation.

```
Rule Nec_copyToRoot
  hasElement w knows variable A
  isLinked w0 w R

  add w0 knows variable A

Rule Nec
  hasElement w knows variable A
  isLinked w u R

  add u variable A

Rule Nec_forReflexivity
  hasElement w knows variable A

  add w variable A
```

Note that the application of the last rule may be restricted to the root node: for the other nodes the first two rules already do the job.

The strategy `S5Strategy` is as follows:

```
Strategy S5Strategy
  repeat
```

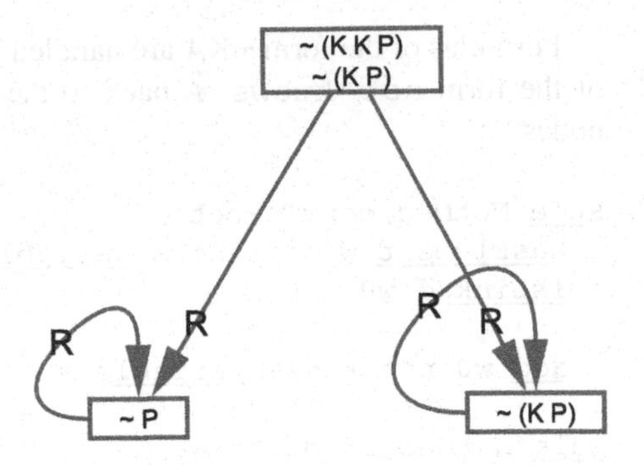

Fig. 4.14 An open premodel for the formula ¬KKP

```
    repeat
      BooleanRules
    end
    Nec_forReflexivity
    Nec_copyToRoot
    Nec
    NotNec_copyToRoot
    NotNec_fromRoot
  end
end
```

If we run it on ¬KKP then we obtain the open premodel that is depicted in Fig. 4.14. Observe that each premodel in the execution trace is a tree of depth one whose children have reflexive edges.

Exercise 52 Replace the rules `Nec` and `Nec_copyToRoot` by two rules where `A` (instead of `knows A`) is respectively added to the root and to all the siblings in the tree. (Call them `Nec_forSymmetry` and `Nec_forEuclideanity`.)

Check that the same models are obtained in this way.

Why is it not a good idea to do the same thing for ¬KA? If you cannot think of a good reason then you should run the resulting strategy on the formula K¬KP. (Better use the step-by-step mode...)

Extracting a Model from an Open S5 Premodel Suppose the tableaux procedure for **S5** stops with an open premodel. As intended, the premodel is a tree of depth at most one whose root is the only reflexive node of the graph. We can transform that premodel into a model by defining the relation R to be the *universal accessibility relation*, that is to say the relation that links each node with each node.

Exercise 53 Prove that the following formulas are valid in **S5**:

1. Axiom **4**: KP → KKP
2. Axiom **5**: ¬KP → K¬KP
3. Axiom **B**: ¬P → K¬KP

Exercise 54 Extend the above tableaux procedure for **S5** so that it does not output a premodel but a legal **S5** model.

Observe that the accessibility relation of the resulting premodels is always universal, i.e., it is an equivalence relation with a single equivalence class. Explain why this is the case. Conclude that the logic **S5** is not only the logic of the class of equivalence relations, but also for the class of universal relations.

Prove formally that for a given model $M = \langle W, R, V \rangle$ with a single equivalence relation R, every pointed model M, w is bisimilar to the pointed model M_w, w such that $M_w = \langle W_w, R_w, V_w \rangle$ with

$$W_w = \{u \mid \langle w, u \rangle \in R(I)\}$$

$$R_w = R \cap (W_w \times W_w)$$

$$V_w(P) = V(P) \cap W_w, \quad \text{for every } w$$

M_w is called the *submodel generated by* w.

Exercise 55 Modify the rules of **S5** in order to obtain a tableaux procedure for the logic of transitive and euclidean relations **K45**. Try to design an explicit procedure right from the start (which is perhaps more natural).

Exercise 56 Modify the tableaux procedure of the logic of knowledge **S5** in order to obtain a procedure for the logic of serial, transitive, and euclidean relations **KD45**, which is the standard logic of belief.

Exercise 57 The multimodal logic $\mathbf{S5}_n$ is popular as the logic of the knowledge of multiple agents.

1. Extend the procedure for **S5** to $\mathbf{S5}_n$.
2. Prove that the formula $\neg(\mathsf{K}_J P \wedge \mathsf{K}_I \neg P)$ is valid in $\mathbf{S5}_n$.
3. Check whether there is an $\mathbf{S5}_n$ model for the following formulas.

   (a) $\mathsf{K}_I P \wedge \mathsf{K}_J \neg P$
   (b) $\mathsf{K}_I(P \wedge \neg \mathsf{K}_J P)$
   (c) $\mathsf{K}_I \neg P \wedge \neg \mathsf{K}_I \neg \mathsf{K}_J P$
   (d) $\mathsf{K}_I \neg \mathsf{K}_J P \wedge \neg \mathsf{K}_I \neg \mathsf{K}_J P$

Exercise 58 Modify the tableaux procedure of $\mathbf{S5}_n$ in order to obtain a procedure for $\mathbf{KD45}_n$.

4.8 HL: Hybrid Logic Nominals

Hybrid logic is a relatively young child in the family of modal logics, as compared to the other logics that we have seen up to now. The language of hybrid logic extends that of modal logic **K** with special atomic formulas N, M, etc. that are called *nominals* or *names for worlds*. Each nominal denotes a given world in the model.

Nominals are important in extensions of the description logic **ALC** that we introduced in Sect. 2.1.5: there, we think of graph nodes as objects of some domain, and nominals allow us to refer to these objects in the language of concepts. Consider a family relationship graph such as that of Fig. 1.15 in Sect. 1.5. Suppose that the language contains nominals for each of the persons. There are in particular the nominals Ann and Cathy, allowing us to express in the language of concepts that the 'Ann node' is linked to the 'Cathy node' by an edge that is labelled isMotherOf.

Language and Semantics of HL The language of basic hybrid logic **HL** is built from a set of atomic formulas $\mathcal{P}$ and a set of nominals $\mathcal{N}$. These two sets have to be disjoint: we suppose $\mathcal{P} \cap \mathcal{N} = \emptyset$.

Kripke models for **HL** have valuation functions $V : \mathcal{N} \cup \mathcal{P} \longrightarrow 2^W$ such that for every $N \in \mathcal{N}$, $V(N)$ is a singleton. In other words, there is exactly one world $w \in W$ such that $w \in V(N)$. A typical validity of **HL** is $(N \land P) \to \Box(N \to P)$.

We refer to [AtC06] for a more detailed introduction to hybrid logic. Now we show how to implement a tableaux procedure for **HL** in LoTREC.

LoTREC Language for HL In LoTREC, we add a unary connective `nominal` in order to distinguish a nominal from an atomic proposition (that is just any word starting by a capital letter). So an expression of the form `nominal N` represents the nominal N. We display `nominal N` as `{N}`, as usually done in the standard description logic syntax for nominals. So the formula

$$\Diamond N \land \Box(p \land \neg N)$$

is written in LoTREC as

```
and pos nominal N nec and P not nominal N
```

and is displayed as

```
<>{N} & [] (P & ~{N})
```

Identifying Nodes with the Same Nominal Let us consider the hybrid logic formula $\neg P \land N \land \Diamond(P \land N)$. With `K_Strategy`—or any other set of rules and strategy that we have introduced up to now—, we obtain the premodel of Fig. 4.15. It is open because the rules for modal logic **K** allow two or more nodes to be contained the same nominal `N`. However, as the latter is a nominal there can only be one world where N dwells, and two graph nodes containing the same nominal `N` should be considered as equal. In particular, they should have the same labels. Therefore each of the nodes that is labelled `N` should also be labelled by both `P` and `~P`. In other words, the premodel is closed, and in consequence the formula will be reported as unsatisfiable.

LoTREC Rules for HL In LoTREC all rules have to be monotonic: there is no way of directly merging two nodes. We therefore proceed as follows.

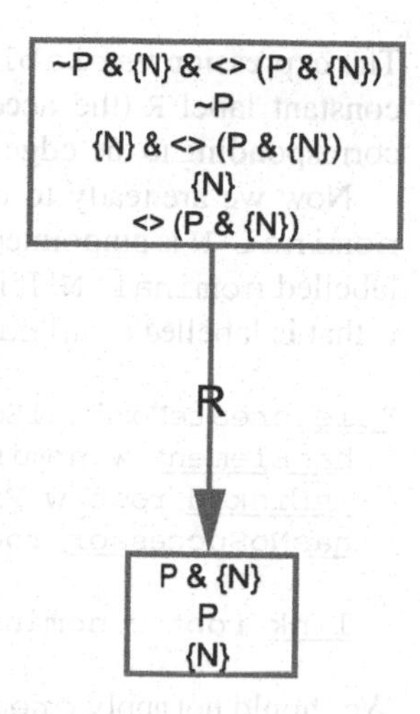

Fig. 4.15 Without node identification, an open premodel for the unsatisfiable formula $\neg P \wedge N \wedge \Diamond(P \wedge N)$

1. The first node containing `nominal N` that is encountered by the procedure will represent all the nodes containing `nominal N`. We call it the `N`-node.
2. We copy all the formulas of a node labelled `nominal N` to that `N`-node.

Practically, we need to pinpoint the `N`-node. To do so we use the root, that is, the first node in the premodel, as a springboard. First, we link the root to all worlds of the premodel by an edge labelled by `FromRoot`. This allows us to retrieve the root at any stage of the model building procedure. Second, we link the root to the unique `N`-node by an edge that is labelled `nominal N`.

We have to identify the root node in some way. We do so by letting it be the only node having a reflexive `FromRoot` edge. (Clearly, the tableaux procedure should ensure that this property is preserved.)

```
Rule DesignateRoot
  isNewNode root

  link root root FromRoot
```

Contrasting with the rules that we have seen up to now, we shall not apply this rule whenever possible: we are going to call it only once, at the beginning of the strategy. (If it was called inside a `repeat` loop it would equip each graph node with a reflexive `R` edge, contrarily to what we want here: to spot a single root node.)

The next rule adds edges from the root to all worlds of the premodel that are labelled `FromRoot`:

```
Rule LinkFromRoot
  isLinked root u FromRoot
  isLinked u w variable L

  link root w FromRoot
```

The expression `variable L` in the rule `LinkFromRoot` may denote either the constant label `R` (the accessibility relation) or a label of the form `nominal N` corresponding to the edge pinpointing the `N`-node that we will introduce below.

Now we are ready to implement Step 1 of our procedure: a node `w` containing `nominal N` is pinpointed as the `N`-node if the root has, as of yet no outgoing edge labelled `nominal N`. It is identified as such by means of an edge from the root to `w` that is labelled `nominal N`.

```
Rule createNominalEdge
  hasElement w nominal variable N
  isLinked root w FromRoot
  hasNoSuccessor root nominal variable N

  link root w nominal variable N
```

We should not apply `createNominalEdge` whenever possible, but only once per `N`. This is going to be ensured in the strategy by means of the `applyOnce` keyword that we have already encountered in the strategy `KD_Strategy` in Sect. 4.5.

Step 2 is implemented by copying into the `N`-node all those formulas that sit at nodes containing `nominal N`. This is achieved by the following rule.

```
Rule copyFormulas
  hasElement u variable A
  hasElement u nominal variable N
  isLinked root u FromRoot
  isLinked root w nominal variable N

  add w variable A
```

In the strategy, the last three rules are called in a <u>`repeat`</u> loop that is preceded by the rule `DesignateRoot`. As we have said, the latter is called only once and initialises the procedure. As we have also said, the existence of at most one `N`-node is ensured prefixing the rule `createNominalEdge` by <u>`applyOnce`</u>. As usual, `BooleanRules` is a sequence of rules for boolean connectives and `Pos`, `NotNec`, `Nec`, and `NotPos` are the rules of `K_Strategy` for modal logic **K**.

```
Strategy Hybrid_Logic_Strategy
  DesignateRoot
  repeat
    LinkFromRoot
    applyOnce createNominalEdge
    copyFormulas
    BooleanRules
    Pos
    NotNec
    Nec
    NotPos
end
```

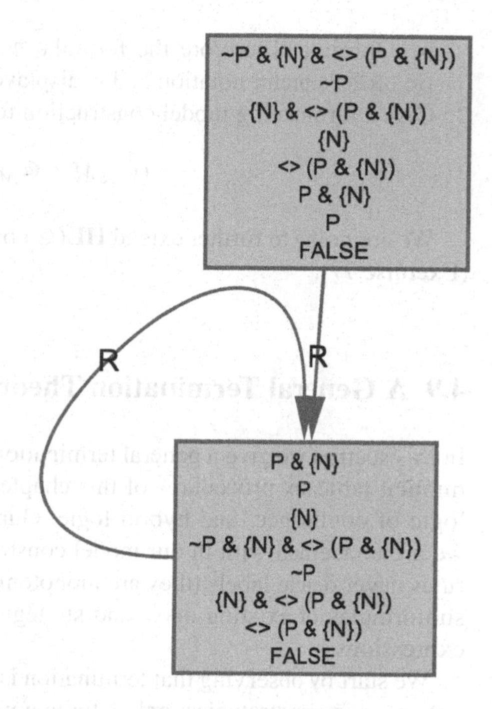

Fig. 4.16 A closed premodel for the unsatisfiable formula $\neg P \land N \land \Diamond(P \land N)$

When we run `Hybrid_Logic_Strategy` on $\Diamond N \land \Box(p \land \neg N)$ then we obtain the closed premodel of Fig. 4.16. In order to transform it into a **HL** model we have to merge the two nodes into a single node.

Exercise 59 Adapt `Hybrid_Logic_Strategy` to the extension of **KT** with nominals.

Exercise 60 The logic **HL**(@) extends **HL** by the special *satisfaction operator* @. The formula $@_N A$ reads "A is true at world N." The set of hybrid formulas $\mathcal{F}or_{\mathbf{HL}(@)}$ is the smallest set such that:

- $\mathcal{P} \cup \mathcal{N} \cup \{\bot\} \subseteq \mathcal{F}or_{\mathbf{HL}(@)}$;
- if $A \in \mathcal{F}or_{\mathbf{HL}(@)}$ then $\neg A \in \mathcal{F}or_{\mathbf{HL}(@)}$;
- if $A, B \in \mathcal{F}or_{\mathbf{HL}(@)}$ then $(A \land B) \in \mathcal{F}or_{\mathbf{HL}(@)}$;
- if $A \in \mathcal{F}or_{\mathbf{HL}(@)}$ then $\Box A \in \mathcal{F}or_{\mathbf{HL}(@)}$;
- if $A \in \mathcal{F}or_{\mathbf{HL}(@)}$ and $N \in \mathcal{I}$ then $@_N A \in \mathcal{F}or_{\mathbf{HL}(@)}$.

The truth condition for the satisfaction operator is:

$$M, w \Vdash @_N A \quad \text{iff there exists } u \in W \text{ such that } u \in V(N) \text{ and } M, u \Vdash A$$

So in order to evaluate a formula $@_N A$ at world w we jump to the unique world u where N is true and check that A is true at u.

1. In LoTREC, extend the language of **HL** by a new binary connective `at`, displayed as `at(_,_)`. Its first argument is a nominal and its second argument

is a formula. Therefore the formula $@_N P$ is written `at nominal N P` in LoTREC's prefix notation and is displayed as `@({N},P)`.
2. Give a terminating model construction for $\mathbf{HL}(@)$ and run it on the formula

$$(@_N M \land @_M P) \to @_N P$$

We are going to further extend $\mathbf{HL}(@)$ by a universal modal operator in Sect. 5.1 (Exercise 77).

4.9 A General Termination Theorem

In this section we give a general termination result. It immediately applies to all the implicit tableaux procedures of this chapter, and it can be adapted to **S5**, $\mathbf{S5}_n$, the logic of confluence, and hybrid logic. Our result exploits the fact that the setting we are concerned with in our model construction methods is very constrained: our rules never delete labels (they are monotonic), they only add formulas that are strict subformulas of existing ones, and strategies have a simple form: they are regular expressions.

We start by observing that termination is trivial when the keyword `repeat` does not occur in our strategies: only a finite number of rules is applied and then the procedure stops. But there is a risk of non-termination as soon as the keyword `repeat` occurs. We give a very simple example to illustrate this: consider

```
Rule AddSuccessor
  isNewNode w

  createNewNode u
  link w u R
```

together with the strategy

```
Strategy Looping_Strategy
  repeat
    AddSuccessor
  end
end
```

This strategy does not terminate, whatever the input formula is: initially, the rule `AddSuccessor` is applicable, creates a new node, and links the root node to it. When `AddSuccessor` is iterated by the strategy then it is again applicable to that new node, creates a third node, and links the former to the latter; and so on. In other words, the strategy tries to build a tree of infinite depth where each node has out-degree one: every node has exactly one successor. (You may try this out yourself with LoTREC; better use the step-by-step mode to avoid loops.)

The above tableaux procedure illustrates that it is not enough if all rules are strictly analytic modulo negation: node creation has to be further constrained. The next exercise shows that it is also edge creation that has to be monitored.

Exercise 61 Augment `K_Strategy` by a rule adding all transitive edges. This is a way of implementing in LoTREC a search only for transitive models (which means that we want to build models of the modal logic **K4**).

Run LoTREC in the step-by-step mode on the formula $\Diamond P \wedge \Box\Diamond P$. Explain why LoTREC loops. What is the form of the graph that such a strategy tries to build?

Exercise 62 Augment `K_Strategy` by the following rule.

```
Rule CreateBrother
  isLinked w u
  hasElement u variable A

  createNewNode v
  add v variable A
```

Run LoTREC in the step-by-step mode on the formula $\Diamond P \wedge \Box\Diamond P$. Explain why LoTREC loops. What kind of graph does such a strategy try to build?

We now give a general termination criterion that covers almost all the implicit procedures in the present chapter. We state it now in a slightly modified form and then sketch its proof.[2] But first of all we have to define some concepts.

4.9.1 Execution Trace

Let $\mathcal{R}$ be a set of LoTREC rules, let S be a strategy combining the rules of $\mathcal{R}$, and let A be a formula. The *execution trace (or computation) of S on* A is the (possibly infinite) sequence of sets of premodels

$$\langle \mathcal{M}_0, \ldots, \mathcal{M}_n, \ldots \rangle$$

that is obtained by the application of S to the initial set of premodels $\mathcal{M}_0 = \{\mathtt{M}_0\}$, where $\mathtt{M}_0 = \langle \{\mathtt{w0}\}, \mathtt{R}_0, \mathtt{V}_0 \rangle$ is such that:

$$\mathtt{R}_0(\mathtt{B}) = \emptyset, \quad \text{for every } \mathtt{B} \text{ in the set of labels } \mathtt{L}$$

$$\mathtt{V}_0(\mathtt{B}) = \begin{cases} \{\mathtt{w0}\} & \text{if } \mathtt{B} = \mathtt{A} \\ \{\emptyset\} & \text{if } \mathtt{B} \neq \mathtt{A} \end{cases}$$

[2] A variant of that theorem was first stated in [GHS06]. However, the condition on node creation (the last condition of Theorem 1) was too weak and is strengthened here.

We say that $M_n \in \mathcal{M}_n$ is *the n-th step* in the execution trace.

For example, the sequence of six sets of premodels (Figs. 3.1–3.6) of Sect. 3.2.2 is the execution trace of `K_Strategy` on `[] P & (<> Q & <>(R v ~P))`.

4.9.2 Monotonic Rules

We have already said in Sect. 3.4 what monotonic tableau rules are: they never delete anything and only add items (i.e., nodes, node labels, edges, and edge labels) to the current premodel. Actually LoTREC's rule language does not contain actions erasing nodes, edges, or labels thereof, and therefore only allows definition of monotonic tableau rules.[3]

Due to monotonicity of rules, for every execution trace $\langle \mathcal{M}_0, \ldots, \mathcal{M}_n, \ldots \rangle$, every integer $n > 0$, and every premodel $M_n \in \mathcal{M}_n$ there is a premodel $M_{n-1} \in \mathcal{M}_{n-1}$ such that M_{n-1} is a *subgraph* of M_n: $W_{n-1} \subseteq W_n$ and for every label $L \in L$, $R_{n-1}(L) \subseteq R_n$ and $V_{n-1}(L) \subseteq V_n$.

Monotonicity allows us to avoid infinite execution traces where e.g. the same edge is added and then removed over and over again: strategies over monotonic rules can only lead to sequences of premodels that are monotonically growing. It remains to find a criterion guaranteeing that they don't grow too much.

4.9.3 Size of Labels

Given a LoTREC label A, the *size* of A, noted lsize(A), is the number of connectives occurring in A.

Here are some examples. The connectives of the label `~~[I] (P & Q)` are the two negations `~`, a necessity operator `[.]`, and a conjunction `&`. We therefore have:

$$\text{lsize}\big(\texttt{\~{}\~{}[I] (P \& Q)}\big) = 4$$

The connectives of the temporal formula `P U (Q v R)` ("`P` until `Q v R`") are the until operator `U` and disjunction `v`. Therefore:

$$\text{lsize}\big(\texttt{P U (Q v R)}\big) = 2$$

This illustrates that complex edge labels also count in the definition. A further illustration is the **PDL** formula `[P ?] Q`. (The formal definition of the language of **PDL** is in Chap. 7.) The modal operator of that formula contains a test of the truth of formula `P`. Its connectives are therefore `[.]` and `?`, and we have lsize(`[P ?] Q`) = 2.

[3] We note in passing that the language of LoTREC contains an unmark action that erases marks and that is therefore not monotonic. However, we have banned it from this book, i.e., we suppose here that unmark is not part of the language.

4.9.4 Sublabels Modulo Negation

We have seen the definition of a subformula in Sect. 2.1.6. For the language of LoTREC we can similarly define the more general notion of a sublabel. Remember that L_0 is the set of all (node or edge) labels.

Given a label A, the set of *sublabels of* A, denoted by SL(A), is inductively defined as:

$$\mathrm{SL}(\texttt{A}) = \{\texttt{A}\} \qquad \text{if } \texttt{A} \in L_0$$

$$\mathrm{SL}(\texttt{c A1 ... An}) = \{\texttt{c A1 ... An}\} \cup \mathrm{SL}(\texttt{A1}) \cup \cdots \cup \mathrm{SL}(\texttt{An}) \qquad \text{else}$$

For example, consider the formula $\langle I\rangle\neg P$, written pos I not P in LoTREC's language and displayed as <I> ~P. In LoTREC's internal notation its sublabels are

$$\mathrm{SL}(\texttt{pos I not P}) = \{\texttt{pos I not P}, \texttt{I}, \texttt{not P}, \texttt{P}\}$$

In the notation as displayed by LoTREC we therefore have

$$\mathrm{SL}(\texttt{<I> ~P}) = \{\texttt{<I> ~P}, \texttt{I}, \texttt{~P}, \texttt{P}\}$$

Every subformula of A is also a sublabel of A, i.e. SF(A) $\subseteq$ SL(A). The converse is not the case because the connectives are not further decomposed in the definition of subformulas. For example, I is a sublabel of <I> ~P while it is not a subformula of $\langle I\rangle\neg P$. To give another example, the set of sublabels of [P ?] Q is

$$\mathrm{SL}\big(\texttt{[P ?] Q}\big) = \big\{\texttt{[P ?] Q}, \texttt{P ?}, \texttt{Q}, \texttt{P}\big\}$$

In the sequel we are always going to suppose that negation—written not in LoTREC and displayed as ~—is a connective of the language. The set of *sublabels of* A *modulo negation* is defined as:

$$\mathrm{SL}^{\pm}(\texttt{A}) = \mathrm{SL}(\texttt{A}) \cup \big\{\texttt{~B} \mid \texttt{B} \in \mathrm{SL}(\texttt{A})\big\}$$

For example, the set of sublabels modulo negation of <I> ~P is:

$$\mathrm{SL}^{\pm}(\texttt{<I> ~P}) = \{\texttt{<I> ~P}, \texttt{I}, \texttt{~P}, \texttt{P}, \texttt{~<I> ~P}, \texttt{~I}, \texttt{~~P}\}$$

We say that B is a *strict sublabel* of A iff B is a sublabel of A different from A itself. In other words, a strict sublabel is an element of SL(A) $\setminus$ {A}. Moreover, B is a *strict sublabel modulo negation of* A when B is a sublabel modulo negation of A and is different from A and ~A, i.e., we require B $\in \mathrm{SL}^{\pm}(\texttt{A}) \setminus \{\texttt{A}, \neg\texttt{A}\}$. Here are some examples:

- ~~P is a strict sublabel modulo negation of <I> ~P;
- ~[P ?] ~Q and ~P are both sublabels modulo negation of [P ?] ~Q;

- Only `~P` is a strict sublabel modulo negation, while `~[P ?] ~Q` is not (while none of `~[P ?] ~Q` and `~P` is a sublabel of `[P ?] ~Q` *tout court*).

Clearly, if B is a sublabel of A then the size of A is at least that of B: if $B \in SL(A)$ then $lsize(B) \leq lsize(A)$. Similarly, if B is a sublabel modulo negation of A then $lsize(B) \leq lsize(A) + 1$.

4.9.5 Strictly Analytic Rules, Connected Rules

All the tableau rules that we have introduced so far have in common that they *only add strict sublabels of the existing labels modulo negation* (except for the rule introducing `False`). We follow Fitting's terminology [Fit83] and call such rules strictly analytic.

Definition 13 (Strictly analytic modulo negation) A LoTREC rule ρ is *strictly analytic modulo negation* if and only if for every label B that is added by ρ (by means of an action of the form add `w B` or link `w u B`) there exists a label A in the condition part of ρ such that

$$B \in \left(SL^{\pm}(A) \setminus \{A, \sim A\}\right) \cup \{\mathtt{False}\}$$

(either B is `False`, or B is a strict sublabel modulo negation of A).

If all rules of $\mathcal{R}$ are analytic modulo negation (and *a fortiori* if they are strictly analytic modulo negation) then the execution trace for A can only have labels from the set $SL^{\pm}(A)$ of sublabels of A (whatever the strategy on $\mathcal{R}$ is).

Lemma 1 *Let $\mathcal{R}$ be a set of rules that is analytic modulo negation. Let S be a strategy on $\mathcal{R}$. Let $\langle \mathcal{M}_0, \ldots, \mathcal{M}_n, \ldots \rangle$ be the execution trace of S on A. Then for every $M_n = \langle W_n, R_n, V_n \rangle \in \mathcal{M}_n$ and for every label B that is not a sublabel of A modulo negation we have $R_n(B) = V_n(B) = \emptyset$.*

As the reader may check, all the implicit LoTREC tableau rules that we have seen so far are strictly analytic modulo negation.

Our last definition is more technical. It is motivated by the fact that termination requires rules which work on patterns of nodes that are all connected in some way.

Definition 14 (Connected rule) A LoTREC rule ρ is *connected* if and only if

1. the node variables and edges that occur in the condition part of ρ make up a connected graph (according to the condition part of ρ),
2. the only conditions occurring in ρ are of one of the following form:
   - <u>isNewNode</u> `w`;
   - <u>hasElement</u> `w A`;
   - <u>hasNotElement</u> `w A`;

- isLinked w u A;
- hasNoSuccessor w.

4.9.6 The Theorem

The following theorem states conditions under which the application of strategy $\mathcal{S}$ to formula A terminates.

Theorem 1 *Let $\mathcal{R}$ be a set of* LoTREC *rules such that*:

1. *every $\rho \in \mathcal{R}$ is monotonic, strictly analytic modulo negation, and connected;*
2. *the only rules that contain the* createNewNode *action or the* link *action have the following form:*

```
Rule nodeAndLinkCreation
  hasElement w B

  createNewNode u
  link w u C
  add u D
```

where B, C, *and* D *may be complex.*

Let $\mathcal{S}$ be any strategy on $\mathcal{R}$. Let A be any input formula. Then the execution trace of $\mathcal{S}$ on A is a finite sequence (of finite sets of finite premodels).

An example of a rule that has the form of nodeAndLinkCreation is the rule Pos of the strategy K_Strategy for modal logic **K** in Chap. 3. This illustrates that rules matched by nodeAndLinkCreation could have any other name. Rule Pos also illustrates that the labels B, C, and D can be of any form, as long as C and D are strict sublabels modulo negation of B (else the rule would not be strictly analytic modulo negation).

All the tableau rules that we have seen in Chaps. 3 and 4 obey the first condition of Theorem 1: they are all monotonic, strictly analytic modulo negation, and connected. Moreover, all the tableaux procedures of Chap. 3 as well as all the tableaux procedures with implicit edges of Sects. 4.1, 4.2, 4.3, 4.4, and 4.5 obey the second condition of Theorem 1. In contrast, all the rules introducing explicit edges violate the second condition of Theorem 1: they have node or edge creating rules that do not have the form of nodeAndLinkCreation. For example, for the tableaux with explicit edges for $\mathbf{K}_2 \oplus$ **Inclusion** of Sect. 4.1 the culprit is the rule Add¬_J_edges_to_I_edges closing under inclusion, and for the explicit tableaux for **KT** of Sect. 4.2 it is the rule AddReflexiveEdges closing under reflexivity. As to the logic of confluence **K.2** of Sect. 4.6, it is the rule Confluence which violates the constraint on node and link creation of Theorem 1; as to the logic **S5** of Sect. 4.7 it is the rule NotNec_fromRoot which violates it. While Theorem 1 therefore does not directly apply to our tableaux procedures for **K.2** and **S5**, one may however establish termination by means of a proof that is very similar to ours below. (In particular, for the case of **S5** the addition of reflexive edges is harmless.)

4.9.7 Proof of the Termination Theorem

We establish the termination theorem by showing that execution traces of strategies whose rules satisfy the conditions of Theorem 1 are finite sequences of premodels each of which is a tree. In a nutshell, our constraints on the form of the tableau rules guarantee that when we execute any strategy combining these rules then (1) we can only yield graphs that are trees; (2) the trees grow monotonically during computation; (3) the *branching factor* of the trees is bounded; and (4) the *depth* of the trees is bounded. It follows that the output trees contain a bounded number of nodes which, in their turn, contain a bounded number of formulas. Before going into the details we have to define trees.

Definition 15 (Ancestor) Given a graph $M = \langle W, R, V\rangle$, a node w is the *ancestor* of a node v if and only if there is a number $n \geq 0$ and a sequence of nodes $\langle v_1, \ldots, v_n\rangle$ such that $v_1 = w$, $v_n = v$, and for every $i < n$, $\langle v_i, v_{i+1}\rangle \in R(I)$ for some $I \in \mathcal{I}$.

The ancestor relation is reflexive: every node w is reachable from w by the empty sequence. It is also transitive.

Definition 16 (Tree, root of a tree) A graph $M = \langle W, R, V\rangle$ is a *tree* if and only if W contains a node $w_0 \in W$, called the *root*, that is an ancestor of every node u in the graph via a unique sequence of edges. The length of that sequence, denoted by $\text{rdist}_M(u)$, is u's *distance from the root.*

For example, the six sets of premodels (Figs. 3.1–3.6) of Sect. 3.2.2 are sets of trees. Moreover, the initial premodel of tableaux procedures is always a tree.

Definition 17 (Depth of a tree) Let $M = \langle W, R, V\rangle$ be a finite tree. The *depth* of M is the greatest distance in that tree:

$$\text{tdepth}(M) = \max_{w \in W} \text{rdist}_M(w)$$

Now we are able to precisely formulate the four arguments that we have given above.

1. Every graph that is generated by the strategy is a tree: this is the case initially; the application of the `nodeAndLinkCreation` rule to a tree yields a new

tree; and the `nodeAndLinkCreation` rule is the only rule adding new nodes or links. Moreover, rules can neither delete edges nor nodes because they are monotonic.
2. Due to monotonicity, the number of labels in a graph can only increase during computation; and due to analyticity, it is bounded by $\text{card}(SL^{\pm}(\texttt{A}))$, i.e. by the cardinality of the set of sublabels modulo negation of the initial formula; thus it cannot change infinitely many times.
3. The width of the trees generated by the strategy is bounded: given an input formula A, no node can have more than $\text{card}(SL^{\pm}(\texttt{A}))$ children. The reason is that for every node w, the rule `nodeAndLinkCreation` creates exactly one successor uB of w per formula labelling w and matched by some B in its condition part; due to analyticity these formulas B are all sublabels modulo negation of the input formula A; therefore there cannot be more such nodes uB than there are elements in $SL^{\pm}(\texttt{A})$.
4. The depth of the trees that are generated by the strategy is bounded. This is for two reasons.
   a. Once a node has been created, its distance from the root will remain the same throughout the execution of the strategy: edges can only be added by the `nodeAndLinkCreation` rule—which for example forbids the addition of transitive edges—and no edges can be removed due to monotonicity.
   b. Due to strict analyticity modulo negation, the farther a node is from the root of the current premodel, the smaller in size are its formulas. Therefore, from some distance to the root, the nodes will be labelled only by atomic formulas; and the nodes that are accessible from there will be empty. However, the `nodeAndLinkCreation` rule cannot apply to an empty node because its condition part requires the existence of a formula labelling that node; and as the `nodeAndLinkCreation` rule is the only rule creating new nodes, none of these empty nodes can have descendants.

We are going to explain these two arguments in more detail below.

In the rest of the proof we establish the last item in more detail. We need the following definition and a lemma.

Definition 18 (Size of a rule) The *size* of a LoTREC rule ρ, denoted by $\text{rsize}(\rho)$, is the number of different node variables occurring in ρ.

The size of a set of rules $\mathcal{R}$ is $\text{rsize}(\mathcal{R}) = \max_{\rho\in\mathcal{R}} \text{rsize}(\rho)$.

For example, the node variables occurring in the Pos rule of K_Strategy in Chap. 3 being w and u, we have $\text{rsize}(\texttt{Pos}) = 2$. For the logic of confluence **K.2**, the size of the Confluence rule is 4.

Definition 19 (Longest label) Given a node w in a premodel M, wsize(w, M) is the length of the longest label decorating either the node w or some outgoing edge of w.

Given a node w and a tree at step n of an execution trace, wsize(w, $\mathtt{M}_n$) is the length of the longest label decorating either the node w or some outgoing edge of w, at step n. Note that due to monotonicity, wsize(w, $\mathtt{M}_n$) can only increase with n.

Lemma 2 *Let* $\langle \mathcal{M}_1, \ldots, \mathcal{M}_n \ldots \rangle$ *be the execution trace of some strategy. For every step n we have:*

- $\text{wsize}(\mathtt{w0}, \mathtt{M}_n) = \text{wsize}(\mathtt{w0}, \mathtt{M}_0)$;
- $\text{wsize}(\mathtt{w}, \mathtt{M}_n) \leq \text{wsize}(\mathtt{w0}, \mathtt{M}_n) - \frac{\text{rdist}_{M_i}(w)}{\text{rsize}(\mathcal{R})}$ *for every node w.*

The first item is the case because of strict analyticity modulo negation: wsize(w0, $\mathtt{M}_n$) cannot change and will remain equal to wsize(w0, $\mathtt{M}_0$).

The second item is an invariant of the execution trace. Its proof is by induction on n.

The induction base is clear because initially all the nodes except w0 are empty.

Let us consider wsize(z, $\mathtt{M}_{n+1}$) and let us suppose that wsize(z, $\mathtt{M}_{n+1}$) > wsize(z, $\mathtt{M}_n$) due to the addition of A (otherwise the inequality is preserved by induction hypothesis). The nodes w and z must be different: otherwise we would have wsize(z, $\mathtt{M}_{n+1}$) > wsize(z, $\mathtt{M}_n$) and we would therefore have added a label to z whose size is greater than that of the existing labels; however, this cannot be the case due to connectedness and strict analyticity modulo negation. The only rules that may increase the world size of some node z from wsize(z, $\mathtt{M}_n$) to wsize(z, $\mathtt{M}_{n+1}$) are those that contain some <u>add</u> z B or <u>link</u> z u B in the action part. Then B must be a strict sublabel of some A occurring in a condition <u>hasElement</u> w A or <u>isLinked</u> w u A. The nodes w and z must be different, and due to connectedness, w is linked to A; this may be via several edges, but at most rsize($\mathcal{R}$) of them.

Let us consider wsize(z, $\mathtt{M}_{n+1}$) and let us suppose that wsize(z, $\mathtt{M}_{n+1}$) > wsize(z, $\mathtt{M}_n$) due to the addition of A (otherwise the inequality is preserved by induction hypothesis). Thus we have:

$$
\begin{aligned}
\text{wsize}(\mathtt{z}, \mathtt{M}_{n+1}) &= \text{lsize}(A) \\
&\leq \text{lsize}(B) - 1 \\
&\leq \text{wsize}(\mathtt{w}, \mathtt{M}_n) - 1 \\
&\leq \text{wsize}(\mathtt{w0}, \mathtt{M}_0) - \frac{\text{rdist}_{M_n}(w)}{\text{rsize}(\mathcal{R})} - 1 \\
&\leq \text{wsize}(\mathtt{w0}, \mathtt{M}_0) - \frac{\text{rdist}_{M_n}(w) + \text{rsize}(\mathcal{R})}{\text{rsize}(\mathcal{R})} \\
&\leq \text{wsize}(\mathtt{w0}, \mathtt{M}_0) - \frac{\text{rdist}_{M_n}(z)}{\text{rsize}(\mathcal{R})}
\end{aligned}
$$

Hence, nodes that are "sufficiently" far from the root will necessarily be empty, and no rule will be further applied on them. Hence the depth of the premodel is bounded by rsize($\mathcal{R}$) $\times$ wsize($\mathtt{w0}, \mathtt{M_0}$). Thus the premodel can only grow (in nodes, edges, and formulas) but is bounded. Hence at some step it stops changing and the computation stops.

This ends the proof of the lemma and of the termination theorem.

Theorem 1 covers the case of tableaux procedures with implicit edges for logics of any combination of the constraints of inclusion, reflexivity, seriality, determinism, and symmetry. This covers logics such as **K**, $\mathbf{K}_n$, $\mathbf{K}_2 \oplus$ **Inclusion**, **KT**, **KD**, $\mathbf{K.Alt}_1$, **KB**, **KDB**, **KTB**.

4.10 Summary

In this chapter we have shown how to extend the basic tableaux procedure of Chap. 3 to take into account some constraints on the accessibility relation. We have studied the case of reflexive, symmetric, linear, serial and confluent accessibility relations, as well as the multimodal case of inclusion of accessibility relations. In addition we considered the case of the logic **S5** and its multimodal version $\mathbf{S5}_n$ whose accessibility relation are equivalence relations. We have given a general termination theorem which covers logics with strictly analytic tableau rules. In the end of the next chapter we shall present a second termination theorem accounting for tableau rules that are analytic, but not strictly so.

Chapter 5
Logics with Transitive Accessibility Relations

Let us pursue the presentation of various classes of Kripke models together with their logics and tableaux procedures. The present chapter is devoted to logics of classes of frames with transitive accessibility relations. The tableaux proof procedures for such logics are a bit more complicated than those that we have seen up to now, and that is why we group them in a separate chapter. Their rules are no longer strictly analytic, which forbids the application of Theorem 1 in order to obtain termination results. Termination can however be ensured by means of loop checking: basically, a node need not be developed further if its labels are contained in one of its ancestor nodes. The implementation of such mechanisms in LoTREC requires new actions, viz. marking nodes, and new conditions, viz. checks of the existence or non-existence of such marks.

We start in Sect. 5.1 with the most basic logic of transitive accessibility relations: **K4**. Its tableaux procedure uses the marking technique of which Sect. 5.2 collects the LoTREC keywords. Section 5.3 is about intuitionistic logic, which is not a modal logic in the strict sense, but is strongly related to the modal logic **KT4**, alias **S4**. Section 5.4 is about Gödel-Löb logic, which is an extension of **K4** by Löb's axiom. We then turn to the formal properties of completeness and termination: we argue for the importance of terminating tableaux methods (Sect. 5.5) and give a general termination theorem for logics with transitive accessibility relations (Theorem 2 of Sect. 5.6). In the end of the chapter we introduce some other modal logics involving transitivity by way of exercises, in particular the extension of **K** by the universal accessibility relation $\mathbf{K}^{[\forall]}$ and hybrid logic with the universal modal connective $\mathbf{HL}(@, \forall)$ (Sect. 5.7).

5.1 K4: Transitivity

K4 is the modal logic of a single transitive accessibility relation. The set of edge labels $\mathcal{I}$ is therefore a singleton. Transitivity is a property of accessibility relations for temporal connectives such as G ("henceforth") and F ("eventually"). A transitive

O. Gasquet et al., *Kripke's Worlds*, Studies in Universal Logic,
DOI 10.1007/978-3-7643-8504-0_5, © Springer Basel AG 2014

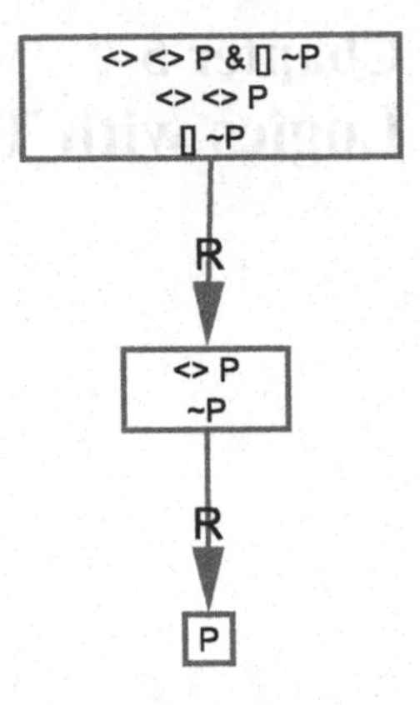

Fig. 5.1 Open premodel for $\Diamond\Diamond P \land \Box\neg P$ obtained by `K_Strategy` of Chap. 3

relation can be characterised by the axiom schema **4**, which may take the form $\Box A \to \Box\Box A$, or alternatively $\Diamond\Diamond A \to \Diamond A$. These axioms are known in epistemic and doxastic logics as the principle of *positive introspection*: if an agent knows that A then he knows that he knows that A. (There is also a principle of *negative introspection* that is expressed in the **5** axiom $\neg\Box A \to \Box\neg\Box A$: if an agent does not know that A then he knows that he does not know that A.)

Figure 5.1 shows an example of an open premodel that is obtained by applying the tableaux procedure of Chap. 3 to the formula $\Diamond\Diamond P \land \Box\neg P$. However, this formula does not have any transitive model. It therefore should not have an open premodel.

Let us turn this open premodel into a transitive model by adding an edge from the first node to the last node: due to the presence of the formula $\Box\neg P$ in the first node, the formula $\neg P$ should in its turn belong to the last node. Such transitive edges can be added straightforwardly by means of the following tableau rule:

```
Rule AddTransitiveEdges
  isLinked w u R
  isLinked u v R

  link w v R
```

We may insert this rule anywhere inside the repeat loop of the `K_Strategy` in Sect. 3.7, for example:

```
Strategy K4_Strategy_With_Transitivity_Naive
  repeat
    BooleanRules
    Pos
    NotNec
    AddTransitiveEdges
    Nec
    NotPos
  end
end
```

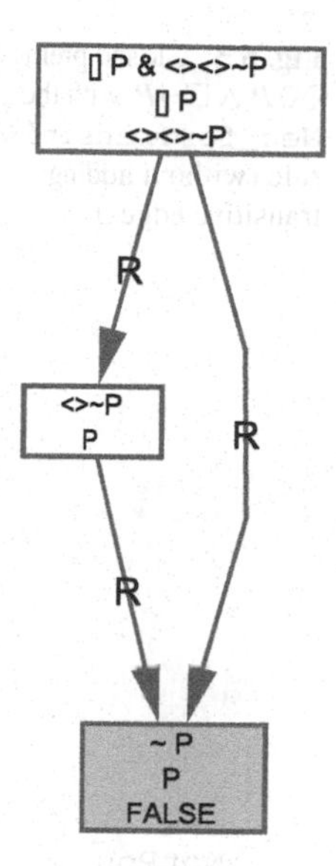

Fig. 5.2 The rule `AddTransitiveEdges` generates a clash

Figure 5.2 shows a closed premodel for the formula $\Diamond\Diamond P \land \Box\neg P$ that is obtained in this way.

The strategy `K4_Strategy_With_Transitivity_Naive` terminates for our example formula. We are going to see below that this is not the case for every input formula. But let us first check the constraints for termination of Theorem 1 of Sect. 4.9 (page 119). While the explicit rule `AddTransitiveEdges` is connected, monotonic and strictly analytic, it violates the last constraint: it adds edges and should therefore have the form of the rule `nodeAndLinkCreation`, but does not.

Adding transitive edges is an explicit way of taking transitivity into account. What about an implicit version of the procedure for **K4**? The presence of the transitive edges can be simulated by propagating the `[]`-formulas in an appropriate way:

```
Rule Nec_forTransitivity
  hasElement w nec variable Formula
  isLinked w u R

  add u variable Formula
  add u nec variable Formula
```

Let us insert `Nec_forTransitivity` and the dual `NotPos_forTransitivity` in the repeat loop of `K_Strategy`:

```
Strategy K4_Strategy_Naive
  repeat
    BooleanRules
    Pos
    NotNec
```

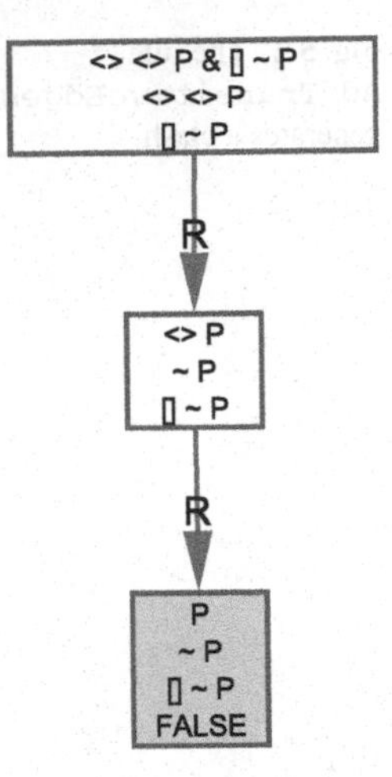

Fig. 5.3 Closed premodel for $\Diamond\Diamond P \wedge \Box\neg P$ with the `Nec_forTransitivity` rule (without adding transitive edges)

```
    Nec
    NotPos
    Nec_forTransitivity
    NotPos_forTransitivity
  end
end
```

Figure 5.3 shows an example of a closed premodel for the formula $\Diamond\Diamond P \wedge \Box\neg P$ that is obtained by the strategy `K4_Strategy_Naive`.

The strategy `K4_Strategy_Naive` terminated for the above example formula, but is there a general result? Well, the rule `Nec_forTransitivity` is analytic and therefore also analytic modulo negation, but not strictly analytic. Hence we cannot apply Theorem 1 of Sect. 4.9 (page 119). Worse, there indeed exist input formulas for which neither of the above two strategies terminates. The simplest such formula is $\Diamond P \wedge \Box\Diamond P$. Take the explicit strategy `K4_Strategy_With_`¬ `Transitivity_Naive`. Figure 5.4 shows the first three premodel sets of the execution trace premodel. They were obtained in the step-by-step mode, with a breakpoint on the successor creating rules `Pos` and `NotNec`. (Remember that between two successive steps all the other rules are applied, in particular the `And`, `Nec` and `AddTransitiveEdges` rules.) Each of the premodel sets in the computation is a singleton. The computation goes as follows:

1. The initial premodel is a single node containing `<> P & [] <> P` (not displayed);
2. The two conjuncts `<> P` and `[] <> P` of the initial formula are added (the first displayed premodel);
3. the `Pos` rule is applied on `<> P`, yielding a first successor that is linked to the root by `R` and that contains `P` (the second displayed premodel);
4. The `Nec` rule adds `<> P` to the first successor (not displayed), which makes the `Pos` rule applicable once again;

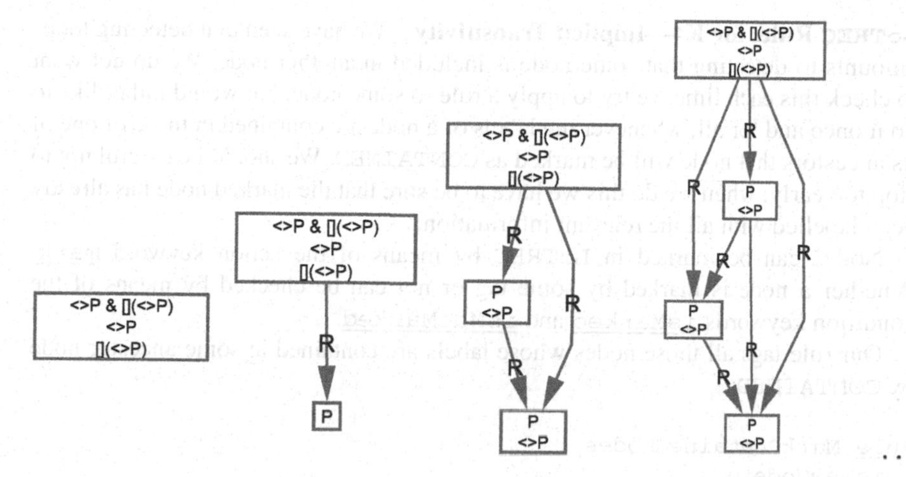

Fig. 5.4 A non-terminating model construction for the formula $\Diamond P \wedge \Box\Diamond P$, with `K4_Strategy_With_Transitive_Naive`

5. The `Pos` rule is applied again, this time to `<> P` in the first successor of the root, yielding a successor of the first successor that is linked to it and that contains `P` (not displayed);
6. the transitive edge from the root to its second successor is added (the third displayed premodel), which makes the rule `Nec` applicable;
7. The `Nec` rule adds `<> P` to the first successor (not displayed), which makes the `Pos` rule applicable once again, and so on: the strategy produces infinitely many successors of the root.

In our example, all the successors of the root contain the same formulas. In the sequel, we define a tableaux method in which we detect such repetitions. In such cases we stop the computation and conclude that the premodel is open. The process of detecting such repeated nodes is called *loop checking*. This detection is only the first stage of the *loop-blocking* process, which consists in blocking the detected looping nodes: no rules can be applied to them any more.

We just saw an example of a loop that cannot lead to a closed premodel. But is this always the case? Is it the case that whenever the construction of a transitive premodel does not terminate then there is such a loop? The answer is yes. Roughly speaking, if rules only add subformulas then a combinatorial argument allows us to limit the number of distinct possible nodes that can be created in an infinite sequence of successors: since there are finitely many subformulas of the input formula and since nodes only contain such subformulas, there can only be finitely many different nodes even in an infinite execution trace. Hence there is at least one node whose labels are repeated. This means that we have a *cycle*, alias a *loop*. These points will be addressed formally at the end of this chapter in terms of a second general termination theorem.

LoTREC Rules for K4—Implicit Transitivity We have seen that detecting loops amounts to detecting that some node is included in another node. We do not want to check this each time we try to apply a rule to some node, but would rather like to do it once and for all: whenever the labels of a node are contained in those of one of its ancestors that node will be marked as CONTAINED. We should be careful not to stop too early: when we do this we have to be sure that the marked node has already been labelled with all the relevant information.

Nodes can be marked in LoTREC by means of the action keyword mark. Whether a node is marked by some tag or not can be checked by means of the condition keywords isMarked and isNotMarked.

Our rule tags all those nodes whose labels are contained in some ancestor node by CONTAINED:

```
Rule MarkContainedNodes
  isNewNode u
  isAncestor w u
  contains w u

  mark u CONTAINED
```

The isAncestor condition of LoTREC matches all those nodes w and u such that w is an ancestor of u via a non-empty sequence of edges, whatever their labels are. So the ancestor relation is irreflexive: u is not an ancestor of u.

The applicability of the Pos rule of K_Strategy should now be restricted by requiring that the node containing the $\Diamond$-formula *should not be blocked*, i.e., it should not be marked as contained in some ancestor node.

```
Rule Pos_ifUnblocked
  hasElement w pos variable A
  isNotMarked w CONTAINED

  createNewNode u
  link w u R
  add u variable A
```

We should order these two rules in the appropriate way: MarkContained-Nodes should be applied before Pos_ifUnblocked and the dual NotNec_ifUnblocked. Moreover, when checking for node inclusion we may make sure that the nodes have been 'filled up' with labels: the boolean rules should have been applied exhaustively. Here is a strategy with implicit edges for **K4**:

```
Strategy K4_Strategy
  repeat
    repeat
      BooleanRules
    end
    MarkContainedNodes
```

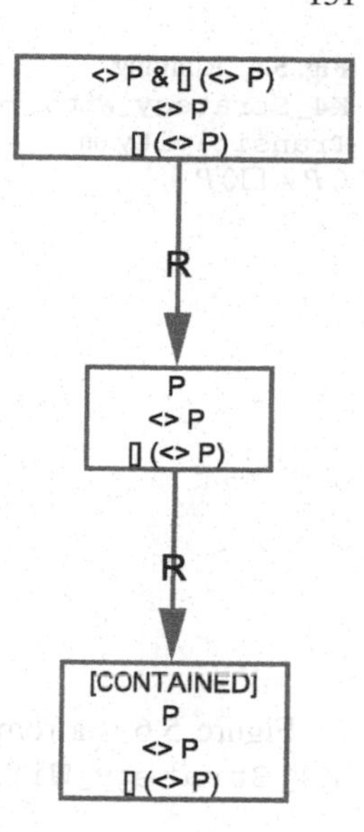

Fig. 5.5 Running `K4_Strategy` on $\Diamond P \wedge \Box\Diamond P$

```
    Pos_ifUnblocked
    NotNec_ifUnblocked
    Nec
    NotPos
    Nec_forTransitivity
    NotPos_forTransitivity
  end
end
```

Figure 5.5 is a remake of the example of Fig. 5.4 and is obtained by running `K4_Strategy` on $\Diamond P \wedge \Box\Diamond P$.

LoTREC Rules for K4—Explicit Transitivity Here is the strategy of the tableaux procedure for **K4** with explicit transitive edges. As before, we start the repeat loop by applying the rules for the boolean connectives as much as possible, iterating `BooleanRules` in a (nested) repeat loop.

```
Strategy K4_Strategy_With_Transitivity
  repeat
    repeat
      BooleanRules
    end
    MarkContainedNodes
    Pos_ifUnblocked
    NotNec_ifUnblocked
    AddTransitiveEdges
    Nec
    NotPos
  end
end
```

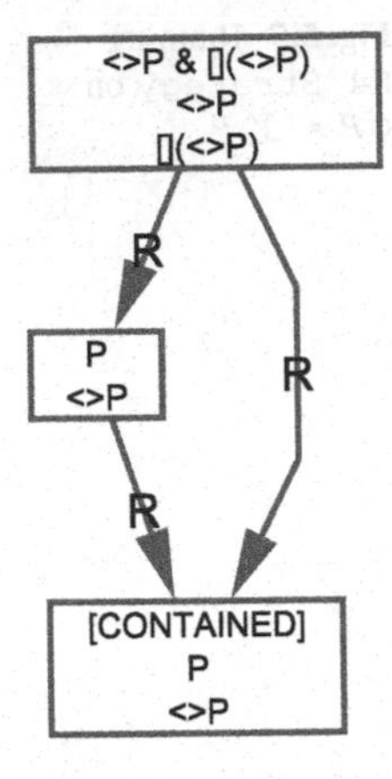

Fig. 5.6 Running `K4_Strategy_With_Transitivity` on $\Diamond P \wedge \Box\Diamond P$

Figure 5.6 is a remake of the example of Fig. 5.4 and is obtained by running `K4_Strategy_With_Transitivity` on $\Diamond P \wedge \Box\Diamond P$.

Extracting a Model from a Saturated Open Premodel Once one has obtained an open premodel like that of Figs. 5.5 and 5.6, one may wish to turn these premodels into **K4** models. While the premodel of Fig. 5.6, is transitive, that of Fig. 5.5 is not. We may apply the rule `AddTransitiveEdges` to establish transitivity. However, the resulting model is not a model of $\Diamond P \wedge \Box\Diamond P$ yet. Indeed, $\Box\Diamond P$ should be true at the root, and therefore $\Diamond P$ should be true at the blocked node, but the latter is not the case: that node does not have any successor.

Let us call `w` the root node and `u` the blocked node that is marked `CONTAINED`. The solution is to make these two nodes identical: while `w` and `u` are already labelled by the same node labels, they should also lead to the same successor nodes. This can be achieved by adding edges from `u` to all those nodes that can be accessed from `w`:

```
Rule CopyOutgoingEdges
  isLinked w v R
  isAncestor w u
  isMarked u CONTAINED

  link u v R
```

In the strategy `K4_Strategy_With_Transitivity`, we have to add the two rules `AddTransitiveEdges` and `CopyOutgoingEdges` after the repeat loop. When run on $\Diamond P \wedge \Box\Diamond P$, the resulting strategy returns the model of Fig. 5.7, which is indeed a transitive model of the input formula.

Exercise 63 Give a terminating tableaux method for the logic of reflexive and transitive frames **KT4**, alias **S4**.

Exercise 64 Give a terminating tableaux method for the logic of serial and transitive frames **KD4**.

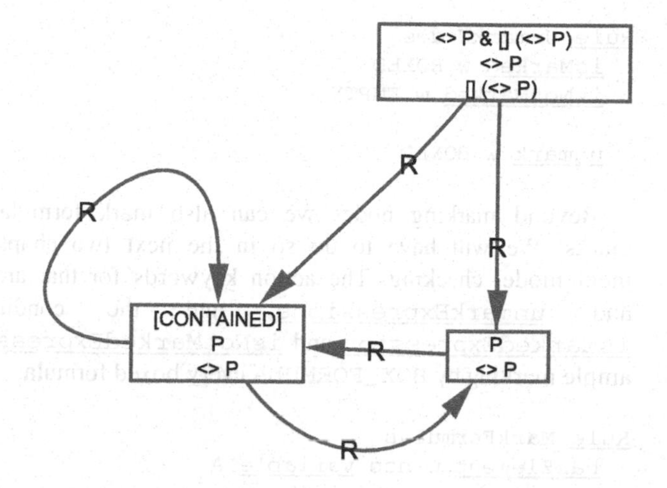

Fig. 5.7 Model obtained after identifying blocked nodes

Exercise 65 Look at the premodels obtained for the formula $\Diamond(P \vee Q) \wedge \Box\Diamond(P \vee Q)$. There may be less of them if the disjunction were not processed in a node that is already contained in some ancestor. Modify `BooleanRules` so that they apply only on formulas of *non-blocked nodes*, and rewrite the strategy in order to check inclusion right after propagation of `[]`-formulas.

5.2 Marking Nodes and Expressions in LoTREC

In this section we interrupt the presentation of various basic modal logics and present the LoTREC actions and conditions for marking and tagging in a grouped way.

There is no predefined tag in LoTREC and we may use instead any other character string. In the rule `MarkContainedNodes` we could have used any other word such as `included` or `nodeToBeBlocked` instead of the tag `CONTAINED`.

Node marks are modified by the action keywords mark and unmark.[1] Whether a node is marked or not can be checked by means of the condition keywords isMarked and isNotMarked. Here is a toy example of a rule marking all nodes containing a boxed formula by the tag `BOXED`:

```
Rule MarkNodes
  hasElement w nec variable A

  mark w BOXED
```

Here is another rule removing this mark from all those nodes marked `BOXED` that are not marked `EMPTY`:

[1]Unmarking actions exist in LoTREC, but no tableaux method in this book uses it.

```
Rule UnmarkNodes
  isMarked w BOXED
  isNotMarked w EMPTY

  unmark w BOXED
```

Beyond marking nodes we can also mark formulas and check for such marks. We will have to do so in the next two chapters in order to implement model checking. The action keywords for that are markExpressions and unmarkExpressions, and the condition keywords are isMarkedExpression and isNotMarkedExpression. Here is a toy example marking by BOX_FORMULA every boxed formula:

```
Rule MarkFormulas
  hasElement w nec variable A

  markExpressions w nec variable A BOX_FORMULA
```

The next rule removes this mark from all formulas that are marked BOX_FORMULA but not DUMB:

```
Rule UnmarkFormulas
  hasElement w nec variable A
  isMarkedExpression w nec variable A BOX_FORMULA
  isNotMarkedExpression w nec variable A DUMB

  unmark w nec variable A BOX_FORMULA
```

Finally, there is a LoTREC condition checking whether *every* R-successor of a node w contains some label A that is marked by M. It takes the form:

```
isMarkedExpressionInAllChildren w A R M
```

and will be used in order to evaluate whether a formula *A* is marked as being true in every successor node.

5.3 Intuitionistic Logic: Reflexivity, Transitivity and Persistence

Almost all logics that we have seen up to now were identified by particular constraints on accessibility relations. The only exception was hybrid logic where special atomic formulas have to be true in exactly one node. We are now going to introduce another logic with a transitive accessibility relation that moreover has a constraint on the valuation function: intuitionistic logic.

Intuitionistic logic has its origins in the philosophy of mathematics: mathematicians such as L.E.J. Brouwer and Arendt Heyting criticised the fact that, in classical propositional logic, principles such as *tertium non datur* ("excluded middle")

$A \vee \neg A$, *reductio ad absurdum* $\neg\neg A \rightarrow A$, or contraposition $(\neg A \rightarrow \neg B) \rightarrow (B \rightarrow A)$ are valid. They argued that this allows for proofs that are *not constructive*. If we rewrite the excluded middle as $A \vee (A \rightarrow \bot)$ then we can see that all these principles are basically about implication.

Let us write $\Rightarrow$ for the intuitionistic implication, which is an implication that is weaker than the material implication $\rightarrow$.[2] In particular, it does not validate the above three principles. In intuitionistic logic, negation is not a primitive connective: intuitionistic negation $\sim A$ is defined to be an abbreviation of $A \Rightarrow \bot$.

Shortly after proposing possible world models, Saul Kripke discovered that these models not only provide a semantics for modal logics, but also for intuitionistic logic: it suffices to interpret the intuitionistic implication $A \Rightarrow B$ in the same way as the so-called *strict implication* $\Box(A \rightarrow B)$. Strict implication is a kind of implication that was at the origin of modern modal logic in the beginning of the 20th century through the work of Lewis and Langford [LL32]. Precisely, the truth condition for intuitionistic implication is:

$$\begin{aligned} M, w \Vdash A \Rightarrow B \quad & \text{iff for every possible world } u \text{ such that } \langle w, u \rangle \in R, \\ & M, u \nVdash A \text{ or } M, u \Vdash B \end{aligned}$$

Kripke found out that the appropriate class of models for intuitionistic logic should have an accessibility relation R that is reflexive, transitive, and *persistent*. (The latter is also called hereditary.)

Definition 20 (Persistence) Let $M = (W, R, V)$ be a Kripke model. M is *persistent* if and only if for every $\langle w, u \rangle \in R$ and every $P \in \mathcal{P}$, if $w \in V(P)$ then $u \in V(P)$.

As the reader may check, none of the formula schemas $A \vee \sim A$, $\sim\sim A \Rightarrow A$, and $(\sim A \Rightarrow \sim B) \Rightarrow (B \Rightarrow A)$ is valid in the class of reflexive, transitive and persistent models.

Intuitionistic logic can be translated into modal logic **S4** via the so-called Gödel translation *tr*:

$$\begin{aligned} tr(P) &= \Box P, \quad \text{for } P \text{ atomic} \\ tr(\bot) &= \bot \\ tr(A \wedge B) &= tr(A) \wedge tr(B) \\ tr(A \vee B) &= tr(A) \vee tr(B) \\ tr(A \Rightarrow B) &= \Box\big(tr(A) \rightarrow tr(B)\big) \end{aligned}$$

[2] 'Being weaker than' has to be understood in the following sense. Let A be a formula of classical propositional logic (written with $\rightarrow$) and let A' be obtained from A by replacing every '$\rightarrow$' by '$\Rightarrow$'. If A' is valid in intuitionistic logic then A should be valid in classical propositional logic, while the converse is not the case: there are formulas A' such that A is valid in classical propositional logic while A' is not valid in intuitionistic logic.

Table 5.1 Definition of the connectives of intuitionistic logic in LoTREC

Connective	Name	Arity	Displayed as	Priority	Associative
$\bot$	`False`	0	`FALSE`		
$\wedge$	`and`	2	`(_ & _)`	3	yes
$\vee$	`or`	2	`(_ v _)`	2	yes
$\Rightarrow$	`iimp`	2	`(_ => _)`	2	no
	`T`	1	`(T _)`	1	
	`F`	1	`(F _)`	1	

It can be proved that an intuitionistic formula A is valid in intuitionistic logic if and only if $tr(A)$ is valid in modal logic **S4** (by induction on the structure of A).

LoTREC Language for Intuitionistic Logic The first thing to do is to introduce the connectives. We collect all the definitions in Table 5.1. The connectives `T` and `F` in the last two lines are auxiliary connectives that are not part of the logical language of intuitionistic logic. Instead, they are required by our tableaux procedure. They will only occur as the outermost connective (which is why we don't care whether they are declared to be associative or not). It is supposed that each formula labelling a node is prefixed by either `T` or `F`. We therefore say that such tableaux are for *signed formulas*: when node `w` has a label "`T A`" then `A` has to be true at `w`, and when `w` has a label "`F A`" then `A` has to be false at `w`.

Note that we could have formulated signed versions of all of our tableaux rules. This is actually often done in the tableaux literature. The advantage of signed tableaux is that we may simplify the definition of a rule being strictly analytic modulo negation (Definition 13): abstracting away from the signs `T` and `F`, we may talk about a rule being strictly analytic *tout court*.

Note that in intuitionistic logic, the label `F A` only requires that `A` is false in the actual world. This contrasts with the label `T A => FALSE` which requires that `A` is false at the actual node *and at all successor nodes*.

LoTREC Rules for Intuitionistic Logic The tableau rule for clashes now has to take the signs `T` and `F` into account.

```
Rule T_F_Clash
  hasElement w T variable A
  hasElement w F variable A

  add w T FALSE
```

The tableau rules for the conjunction $\wedge$ are basically the same as in classical propositional logic. They however have to take the signs `T` and `F` into account.

```
Rule T_And
  hasElement w T and variable A variable B
```

```
  add w T variable A
  add w T variable B

Rule F_And
  hasElement w F and variable A variable B

  duplicate premodel_copy
  add w F variable A
  add premodel_copy.w F variable B
```

The tableau rules for disjunction ∨ are symmetric to those for conjunction:

```
Rule F_Or
  hasElement w F or variable A variable B

  add w F variable A
  add w F variable B

Rule T_Or
  hasElement w T or variable A variable B

  duplicate premodel_copy
  add w T variable A
  add premodel_copy.w T variable B
```

The intuitionistic implication behaves like a connective of the ◇ kind when we require it to be false, i.e., when it is signed `F`: it is here where we have to check for loops.

```
Rule F_IImp
  hasElement w not iimp variable A variable B
  isNotMarked w CONTAINED

  createNewNode u
  link w u R
  add u T variable A
  add u F variable B
```

If an intuitionistic implication `A => B` is signed "`T`" then it behaves like a □ connective. Remember that we have to take into account that the accessibility relation is reflexive and transitive. We do so in an implicit way, adapting the rules for the logics **KT** and **K4**. So there are two tableau rules for `T(A => B)`: the first amounts to adding the material implication `A -> B` to the actual node, and the second adds the intuitionistic implication `A => B` to all accessible nodes.

```
Rule T_IImp_forReflexivity
  hasElement w T iimp variable A variable B
```

```
  duplicate premodel_copy
  add w F variable A
  add premodel_copy.w T variable B

Rule T_IImp_forTransitivity
  hasElement w T iimp variable A variable B
  isLinked w u R

  add u T iimp variable A variable B
```

Rule `F_IImp` checks for the `CONTAINED` tag. In order to generate such tags we take over the rule `MarkContainedNodes` from the logic **K4**.

```
Rule MarkContainedNodes
  isNewNode u
  isAncestor w u
  contains w u

  mark u CONTAINED
```

Finally, we have not yet taken persistence into account. We do so in a straightforward way by copying along the accessibility relation all the atomic formulas that are marked true. This requires the LoTREC keyword `isAtomic`: the condition `isAtomic A` checks whether `A` is atomic or not. (More precisely, it checks at runtime whether the label that is matched by `A` is atomic or not.)

```
Rule AtomicPersistence
  hasElement w T variable A
  isAtomic variable A
  isLinked w u

  add u T variable A
```

We note in passing that there also exists a LoTREC condition `isNotAtomic`. It is not used in this book.

Let us put these tableau rules together in a strategy. That strategy works as follows: we start by saturating each node by means of the rules for $\wedge$ and $\vee$ and the rule `T_IImp_forReflexivity` (which saturates a single node in the same way as the rules for the boolean connectives do). Thereafter we check for inclusion in some ancestor, in which case the node is blocked. For unblocked nodes, new successor nodes are created by rule `F_IImp` and are then 'filled up' by the rules `T_IImp_forTransitivity` and `AtomicPersistence`.

```
Strategy LJ_Strategy
  repeat
    repeat
```

```
            T_F_Clash
            T_And
            F_Or
            T_Or
            F_And
            T_IImp_forReflexivity
        end
        MarkContainedNodes
        F_IImp
        T_IImp_forTransitivity
        AtomicPersistence
    end
end
```

Exercise 66 Run `LJ_Strategy` on the following formulas.

1. $P \vee (P \Rightarrow \bot)$
2. $((P \Rightarrow \bot) \Rightarrow \bot) \Rightarrow P$
3. $P \Rightarrow ((P \Rightarrow \bot) \Rightarrow \bot)$
4. $((P \Rightarrow Q) \Rightarrow P) \Rightarrow P$

5.4 GL: Transitivity and Noetherianity

Noetherianity means that there is no infinite chain: a relation R is *noetherian* if and only if there is no infinite sequence of elements $w_0, w_1, w_2, \ldots$ such that $\langle w_0, w_1 \rangle \in R$, $\langle w_1, w_2 \rangle \in R$, $\langle w_2, w_3 \rangle \in R$, etc. Gödel-Löb logic **GL** is the monomodal logic of transitive and noetherian frames.

GL is important in mathematical logic as the logic of provability in arithmetic. Let us make this more precise. (We refer the reader to [Ver10] for more detailed explanations.) Let us write Φ, Ψ, etc. for statements in the language of Peano arithmetic, such as $2^2 = 4$ (which is provable) or $1 + 1 = 3$ (which is inconsistent). The Gödel number of Ψ, usually written $\ulcorner\Psi\urcorner$, is the result of assigning a numerical code to Ψ, as done in Gödel's proof of the incompleteness of arithmetic. The provability predicate $\mathsf{Pr}(\ulcorner\Psi\urcorner)$ has as an argument the Gödel number of some statement Ψ and is itself of the form $\exists n\, \mathsf{Proof}(n, \ulcorner\Psi\urcorner)$, where $\mathsf{Proof}(n, \ulcorner\Psi\urcorner)$ means that the Gödel number n codes a proof of the statement Ψ from the axioms of Peano arithmetic. An example of a statement involving Pr is

$$\Phi = \Psi \wedge \neg\mathsf{Pr}(\ulcorner\Psi\urcorner) \wedge \neg\mathsf{Pr}(\ulcorner\neg\Psi\urcorner)$$

The formula Φ says: Ψ is true but not provable, and its negation is not provable either. In other words: Ψ is true, but both $\neg\Psi$ and Ψ are consistent.

An *interpretation* (also called a *realisation*) is a function f assigning to every atomic formula P of the language of **GL** some sentence in arithmetic $f(P)$. That interpretation is extended to the whole language of **GL** by stipulating the following:

- $f(\bot) = \bot$;
- $f(\neg A) = \neg f(A)$;
- $f(A \wedge B) = f(A) \wedge f(B)$;
- $f(A \vee B) = f(A) \vee f(B)$;
- $f(\Box A) = \mathsf{Pr}(\ulcorner f(A) \urcorner)$.

According to Solovay's theorem [Sol76], for every formula A of the language of **GL** it is the case that A is a theorem of **GL** if and only if for every interpretation f, $f(A)$ is provable in arithmetic. This entitles us to say that the modal connective $\Box$ of **GL** captures the properties of the provability predicate Pr of arithmetic.

The logic **GL** can be axiomatised by extending the axiomatisation of the logic of transitive frames **K4** with Löb's axiom schema **L**: $\Box(\Box A \rightarrow A) \rightarrow \Box A$. The latter can also be formulated as $\Diamond A \rightarrow \Diamond(A \wedge \neg\Diamond A)$, which perhaps better explains the relation with noetherianity: if a world where A is true is accessible then the last such world is also accessible, where 'the last such world' means that A is false from there on.

Exercise 67 Prove the axiom schema **4**: $\Box A \rightarrow \Box\Box A$ in the axiom system $\mathbf{K} \oplus \mathbf{L}$, i.e., **K** extended by Löb's axiom schema **L**. (Hint: in **L**, replace A by $B \wedge \Box B$ and prove that the antecedent is implied by $\Box A$.)

LoTREC Rules for GL—Implicit Noetherianity Let us reformulate the truth condition for the $\Diamond$ connective in the light of the version $\Diamond A \rightarrow \Diamond(A \wedge \neg\Diamond A)$ of axiom schema **L**:

$$M, w \Vdash \Diamond A \quad \text{iff there is a } u \text{ such that } \langle w, u \rangle \in R \text{ and } M, u \Vdash A \wedge \neg\Diamond A$$

The tableau rule for the $\Diamond$ connective translates this into LoTREC:

```
Rule Pos_forNoether
  hasElement w pos variable A

  createNewNode u
  link w u R
  add u variable A
  add u not pos variable A
```

We replace the Pos rule of the naive strategy K4_Strategy_Naive for **K4** of Sect. 5.1 by the rule Pos_forNoether. Call the resulting strategy GL_Strategy. The reader may expect that strategy may loop; however, this is not the case: the addition of not pos variable A also does the job of blocking loops! In other words: although **GL** has a transitive accessibility relation, we do not need inclusion tests and node blocking. The reader may check this by running some examples.

Exercise 68 Run GL_Strategy on $\Diamond P \wedge \Box\Diamond P$ and $\Box\Diamond P$. Explain why the premodel for $\Box\Diamond P$ is open.

Exercise 69 Prove that the strategy `GL_Strategy` never loops. (Hint: suppose the contrary.)

Exercise 70 Consider the class of transitive, reflexive, and noetherian frames.

1. Show that it is empty.
2. Augment `GL_Strategy` by one of the rules for reflexivity of Sect. 4.2 (either the explicit `AddReflexiveEdges` or the implicit `Nec_forReflexivity`). Run it on the formula $\Diamond P$ and observe that you obtain a closed premodel. Observe that it does not close when run on the formula P. What do you conclude about completeness of your strategy w.r.t. the empty set of frames?
3. Perform a similar analysis of the class of transitive, serial, and noetherian frames.

Exercise 71 In the previous exercise we have seen that the class of reflexive, transitive and noetherian frames is empty. Let us now modify slightly the definition of noetherianity as follows. Call a relation *strictly noetherian* if and only if there is no infinite chain except chains built from reflexive loops. Formally, a relation R is strictly noetherian if and only if there is no infinite sequence $w_0, w_1, w_2, \ldots$ such that $\langle w_0, w_1 \rangle \in R$, $\langle w_1, w_2 \rangle \in R$, $\langle w_2, w_3 \rangle \in R, \ldots$ and such that $w_0 \neq w_1$, $w_1 \neq w_2, \ldots$. The logic of reflexive, transitive and strictly noetherian frames is called **Grz**.

Implement a model construction method for **Grz**.

5.5 Completeness vs. Termination

In general, soundness proofs are straightforward and completeness proofs are involved. But there exists a standard methodology for completeness proofs as exposed e.g. in [Fit83, Gor99]. It is usually based on a *fair strategy* giving equal rights to all the rules: every applicable rule $\rho \in \mathcal{R}$ will eventually be applied. For example, all the strategies of Chaps. 3 and 4 are fair: they loop through the rules and apply each rule to every premodel in the set of current premodels. Fair strategies are of fundamental importance in the tableaux method. Indeed, if it may happen that some rule is applicable but never applied then we may fail to produce enough information to detect a contradiction and end up with some open premodel that pretends to be saturated but to which no model can be associated.

Fair algorithms do not always terminate. We have seen in Sect. 5.1 that the fair application of the tableau rules for **K4** to the formula $\Diamond P \wedge \Box\Diamond P$ runs forever, as illustrated in Fig. 5.4 for an explicit procedure. Therefore, standard presentations of the tableaux method of the literature such as [Fit83, Gor99, Mas00] have a last part with an algorithm combining the rules in a way such that termination is guaranteed. That algorithm typically goes as follows.

- Some rules are only applied once. This is typically the case for the reduction rules for $\neg\neg$, $\wedge$, and $\neg\vee$ (called "α-rules" in Fitting's terminology), for the branching

rules for $\vee$ and $\neg\wedge$ ("β-rules"), and for the node-creating rules for $\Diamond$ and $\neg\Box$ ("π-rules"). In contrast, the 'node-filling' rules for $\Box$ and $\neg\Diamond$ ("ν-rules") may have to be applied more than once in order to propagate $\Box$ formulas to newly created nodes.
- The application of the node-creating π-rules is blocked for those nodes that are included in—or identical to—some ancestor node.

Then a combinatorial argument limits the number of possible nodes that can be created in a graph as well as the number of possible graphs.

As stressed in [FdCGHS05], for such a modified algorithm *completeness has to be re-proved.* In the literature often only brief and informal arguments are employed which claim that the restriction on rule application that is imposed by the terminating algorithm is harmless in what concerns completeness.

It is clear that proving completeness for the fair strategy is already a hard task. To prove completeness for the terminating strategy is even harder. Moreover, when working with complex modal logics one often varies both the rules and the strategy. For instance, in a logic of action and knowledge one may wish to 'plug in' the axiom $\mathsf{K}_I[a]A \to [a]\mathsf{K}_I A$ which says that if agent I knows that A is the case after action a then after a, I knows that A. This principle is called *no forgetting* or *perfect recall* in the literature. In LoTREC, this can be implemented in an explicit way by adding a rule ρ which does the following: whenever there are nodes `w`, `u`, `v` such that $\langle$`w, u`$\rangle \in$ `R(a)` and $\langle$`u, v`$\rangle \in$ `R(I)` then create a new node `t`, add an edge labelled `I` from `w` to `t` and add an edge labelled `a` from `t` to `v`. This modifies the set of tableau rules $\mathcal{R}$. It also requires a modification of the strategy because we have to insert that rule somewhere. When doing so we may try out different places for that rule in order to avoid loops or in order to obtain smaller premodels. It is time-consuming and often difficult to re-prove both completeness and termination for each of these modifications.

We believe that in practice, formal termination proofs are more important than formal completeness proofs: it is sufficient to only *conjecture completeness*. To support our claim, suppose we have at our disposal a tableaux procedure for logic Λ that is implemented in LoTREC and that is both terminating and sound, together with a procedure that allows us to build a model from an open premodel (which typically consists in closing the accessibility relation under some property). Suppose we want to know whether a given formula A is Λ-satisfiable or not. We proceed as follows:

- If the procedure returns a set of premodels all of which are closed then soundness of the algorithm guarantees that A is Λ-unsatisfiable.
- If the procedure returns at least one open `M` with root `w0` then according to our completeness conjecture we should be able to turn it into a Λ-model of A. It remains to verify whether this is indeed the case or not. This is done in the following steps.

  1. Transform the premodel into a Kripke model M with actual world w_0;
  2. Check whether M is a Λ-model;
  3. Check whether $M, w_0 \Vdash A$;

4. If $M, w_0 \Vdash A$ then A is indeed Λ-satisfiable.
5. If $M, w_0 \nVdash A$ then try to transform another open premodel into a Λ-model and check whether A is true at the root; if this fails for all of the open premodels then we have discovered that our completeness conjecture was erroneous and that the tableaux procedure has to be modified to 'get closer to completeness.'

Our argumentation can be summarised as a justification of procrastinating a completeness proof: for a given logic Λ and a sound and terminating tableaux procedure, there is no need to bother with completeness as long as the premodels returned by the procedure can be turned into Λ-models of the input formula. Moreover, this can be checked easily. It is clearly interesting to have a complete tableaux procedure because it avoids continuous amending of the procedure; however, completeness is not as vital a property as are soundness and termination.

5.6 Another General Termination Theorem by Checking for Loops

In our first termination theorem of the previous chapter (Theorem 1, page 119), one of the conditions for termination of the tableaux procedure was that the tableau rules must be strictly analytic. The tableaux procedures of the present chapter violate this constraint and we had to insert loop tests and blocking in order to prevent them from running forever. The following theorem gives conditions under which this ensures termination. It was first proved in [GHS06].

Theorem 2 ([GHS06]) *Let $\mathcal{R}$ be a set of rules such that:*

- *for every $\rho \in \mathcal{R}$, ρ is monotonic;*
- *for every $\rho \subset \mathcal{R}$, ρ is analytic: for all labels* `C` *added by ρ there exists a label* `B` *in the condition part of ρ such that* `C` $\in \mathrm{SL}^{\pm}($`B`$) \setminus \{$`B, ~B`$\} \cup \{$`FALSE`$\}$, *i.e.,* `C` *is a non-strict subformula of* `B` *modulo negation;*
- *$\mathcal{R}$ contains the following rule:*

```
Rule MarkContainedNodes
  isNewNode u
  isAncestor w u
  contains w u

  mark u CONTAINED
```

- *the only rules that contain* `createNewNode` *are of the following form:*

```
Rule Pos_ifUnblocked
  hasElement w B
  isNotMarked w CONTAINED
```

```
createNewNode z
link w z R
add z C
```

Let $\mathcal{S}$ be any strategy on $\mathcal{R}$ such that all node-creating rules are immediately preceded by the `MarkContainedNodes` *rule. Then the application of $\mathcal{S}$ to any input formula* `A` *terminates.*

Proof First, the following holds for the execution trace:

- Due to monotonicity, rule application can only add nodes and labels, but may never delete anything.
- Due to analyticity of rules, every node contains at most a bounded number of formulas; in particular, nodes and edges have at most one occurrence of the same label.
- Due to the restricted form of rules containing `createNewNode`, both the depth and the breadth of the premodels are bounded.

In the rest of the proof we focus on the last point: just as in the proof of Theorem 1, we prove that (1) application of $\mathcal{S}$ can never lead to premodels of infinite depth, and (2) application of $\mathcal{S}$ can never lead to premodels with infinite branching factor.

The argument for (1) is that creation of new nodes is conditioned by non-inclusion in an ancestor node, which—due to the precedence condition on $\mathcal{S}$—is always tested before node creation. Therefore a branch of infinite length would contain an infinite number of nodes having different associated formula sets. This cannot be the case because our rules are monotonic: no formulas are erased. This also gives us an upper bound: the length of branches is bounded by an exponential in the length of the input formula `A`. More precisely, the length of branches in an execution trace is bounded by $2^{\mathrm{card}(\mathrm{SL}(\mathtt{A}))} \leq 2^{\mathrm{length}(\mathtt{A})}$ (cf. Sect. 2.1.5).

As to (2), infinite branching at some node `w` could only be produced by creating infinitely many new nodes `u` that are linked to `w`. According to our conditions on $\mathcal{R}$, nodes can only be created by rules of the form of the `Pos_ifUnblocked` rule. Let us denote these rules by $\rho(\mathtt{B}, \mathtt{C}, \mathtt{R})$, where the parameter `B` is the node label in the condition part, `C` is the node label in the action part, and `R` is the edge label. There are at most $\mathrm{card}(\mathcal{R}) - 1$ such rules. Moreover, due to analyticity the number of formulas of `w` matching `B` is bounded by the number of sublabels of the input formula `A`, which is bound in the length of `A` (see Sect. 2.1.5). Therefore there can at most be $(\mathrm{card}(\mathcal{R}) - 1) \times \mathrm{length}(\mathtt{A})$ successors of `w`: the branching factor is bounded.

This ends the proof. □

This theorem covers tableaux methods for modal logics that are axiomatised by the standard axioms **T**, **B**, **4**, **5**. It moreover applies to intuitionistic logic and to hybrid logic **HL**$(@, \forall)$.

To sum it up, Theorem 1 of Chap. 4 (page 119) and the above Theorem 2 give us very general termination criteria for `LoTREC` strategies. Together, they cover the standard modal logics, including linear-time temporal logic **LTL** and propositional dynamic logic **PDL** that we shall introduce in the next chapter. For logics that do not fit into the above framework termination has to be proved case by case.

5.7 Other Logics as Exercises

Exercise 72 Give a terminating model construction method for the fusion of **K** and **S4**, i.e. for the bimodal logic **S4**$\otimes$**K**.

Exercise 73 Consider the logic (**S4**$\otimes$**K**) $\oplus$ **Inclusion** i.e., a bimodal logic with two modal connectives, the first being interpreted by a transitive accessibility relation that is included in the accessibility relation for the second modal connective.
Give a terminating model construction method for that logic.

Exercise 74 Consider a bimodal logic with an axiom of permutation: (**K**$\otimes$**K**) $\oplus$ $\langle I\rangle\langle J\rangle A \to \langle J\rangle\langle I\rangle A$.
Give a model construction method for it. (Hint: it is not so easy to ensure termination.)

Exercise 75 Consider (**K**$\otimes$**K4**) $\oplus$ **Inclusion**, i.e., the logic of the class of frames with two accessibility relations $R(I)$ and $R(J)$ such that $R(J)$ is transitive and such that $R(I)$ is contained in $R(J)$.
Give a terminating model construction method for that logic. (Hint: reconsult Sect. 4.1 on the inclusion constraint, in particular Exercise 43 on page 93.)

Exercise 76 $\mathbf{K}^{[\forall]}$ is a logic with two modalities: a **K**-modality $\Box$ and an **S5**-modality $[\forall]$, called the *universal modality*. The models of this logic have a *universal accessibility relation* for $[\forall]$. This means that the formula $[\forall]A$ is interpreted as follows:

$$M, w \Vdash [\forall]A \quad \text{iff } M, u \Vdash A \text{ for every } u \in W$$

The logic can be axiomatised by the axioms and inference rules for **K** for $\Box$, the axioms and inference rules for **S5** for $\forall$, plus the inclusion axiom $[\forall]P \to \Box P$.
Define $[\forall]$ in LoTREC as `univ` and display it as `U`.
Give a terminating model construction for this logic.

Exercise 77 **HL**$(@, \forall)$ extends **HL**$(@)$ of Exercise 60 in Sect. 4.8 by the universal modal connective $[\forall]$ of the previous exercise. Together with the nominals of **HL**$(@)$, the @ connective allows us to say things that we cannot express in the standard modal language: we can represent with hybrid formulas some frame constraints that were not definable up to now, such as irreflexivity, asymmetry and antisymmetry. Here is an example: the formula $@_N\Box\neg N$ says that the world named N is not reachable from itself by the accessibility relation. Thus this formula is valid on precisely those frames which are irreflexive: such frames can be characterized by the axiom schema $@_N\Box\neg N$. We refer the reader to [AtC06] for more material on the issue.

1. Prove that the formula $@_N A \leftrightarrow [\forall](N \rightarrow A)$ is valid. What do you conclude about the minimal set of connectives for **HL**$(@, \forall)$?
2. Give a terminating model construction for **HL**$(@, \forall)$.

5.8 Summary

In this chapter we have explained the model construction method for logics with transitive accessibility relations, such as **K4**, **S4**, and intuitionistic logic. For these logics, termination has in general to be enforced by blocking nodes that are included in ancestors.

Chapter 6
Model Checking

Let us recall the definition of the model checking problem of Sect. 2.5:

- Input: a finite Kripke model M, a world w of M, a formula A;
- Output: "Yes" if $M, w \Vdash A$; "No" otherwise.

In the present chapter we present a procedure implementing model checking in LoTREC. We start by an example where we perform the check by hand (Sect. 6.1) and then show how to implement the procedure with LoTREC (Sect. 6.2). Beyond the marking actions and marking conditions that we have already put to work in the previous chapter in order to check for loops, this requires the new condition keyword isMarkedExpressionInAllChildren which allows to check the presence of some mark in all children of a node.

6.1 Model Checking by Hand

In order to decide whether A is true in world w of model M we proceed as follows.

1. Initialisation: we write down the pointed model $\langle M, w\rangle$ and the formula A.
2. Top-down questions: we analyse the input formula by recursively asking questions about truth of its subformulas.
3. Bottom-up answers: we synthesise the answers about the truth of subformulas.

In the following subsections we detail each phase by means of an example.

Initialization Phase We have to enter the input of the model checking problem: the pointed model $\langle M, w\rangle$ and the formula A to be checked. For example, given the set of atomic formulas $\mathcal{P} = \{P, Q\}$, let us consider the following finite Kripke model $M = (W, R, V)$ where:

$$W = \{w, u, v, x\}$$
$$R = \big\{(w, u), (w, v), (v, u), (v, x)\big\}$$

O. Gasquet et al., *Kripke's Worlds*, Studies in Universal Logic,
DOI 10.1007/978-3-7643-8504-0_6, © Springer Basel AG 2014

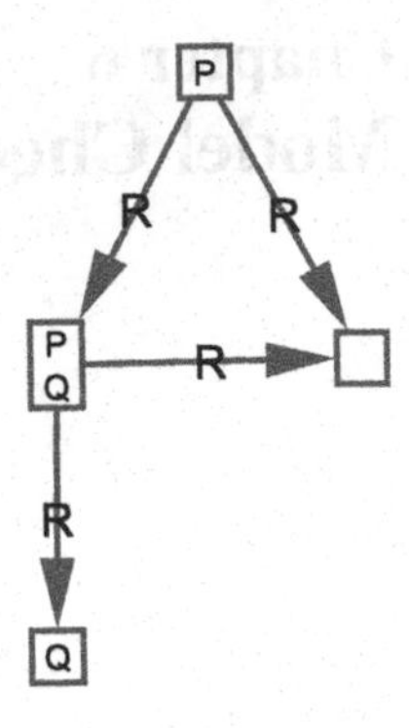

Fig. 6.1 Example of a model

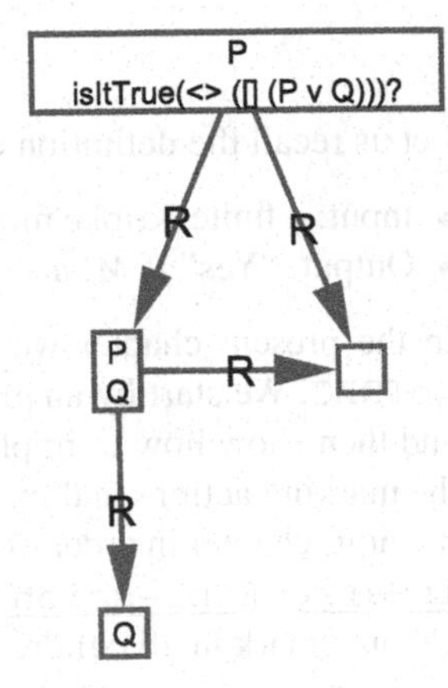

Fig. 6.2 Input of a model checking instance

$$V(P) = \{w, v\}$$
$$V(Q) = \{v, x\}$$

Figure 6.1 shows a graphical representation of M where w is the node at the top of the graph, u is at the right, x is at the bottom, and v is the fourth node. Remember that an atomic proposition that does not appear in a node is considered to be false at that node.

Consider the formula $A = \Diamond\Box(P \vee Q)$: is A true in the top world of the model M of Fig. 6.1, i.e., do we have $M, w \Vdash A$? We would like to write down this question in our graph. We should not add A to the actual world w because this would beg the question. We rather add to w an expression that stands for the question "Is A true at w?" Figure 6.2 shows how the input of our model checking problem is represented in LoTREC.

Top-Down Questions Phase Unless A is an atomic formula, i.e., an element of $\mathcal{P}$, we are not immediately able to answer the question "Is A true at w?" We have to break up A into simpler subformulas and ask questions about these subformulas. For example, according to the truth conditions $\Diamond\Box(P \vee Q)$ is true at w if and only

Fig. 6.3 Top-down questions phase, step 1

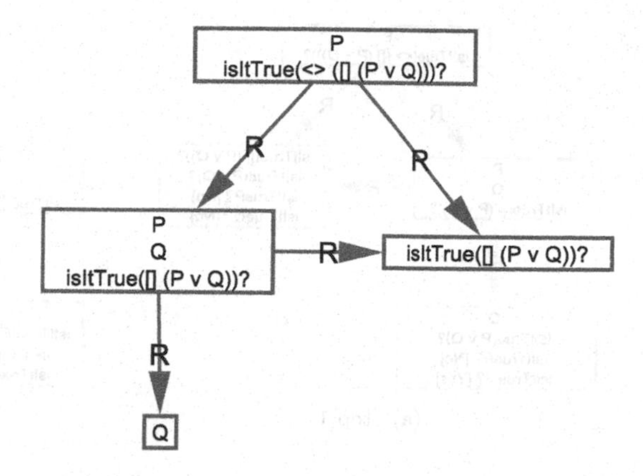

if its subformula $\Box(P \vee Q)$ is true at least at one of the successors of w, i.e., at u or at v. That is why in Fig. 6.3, we add questions about the formula $\Box(P \vee Q)$ to both the right successor u and the left successor v of w.

Now $\Box(P \vee Q)$ is true at the left successor v of w if and only if the formula $P \vee Q$ is true at all successors of v. That is why we add questions about the truth of $P \vee Q$ to all the successors u and x of v, as shown in step 2 of Fig. 6.4.

This newly added formula $P \vee Q$ is true at u (respectively x) if and only if one of its disjuncts P or Q is true at the same world u (respectively x). This is why we add questions about these disjuncts to the corresponding worlds u and x in step 3 of Fig. 6.4. These disjuncts being atomic, no further decomposition is needed, and the answers about the truth values can be synthesised from now on.

We may halt the top-down questions phase when we ask for the truth of $\Diamond$-formulas or $\Box$-formulas in leaf nodes (nodes with no successors). For example, at u the formula $\Box(P \vee Q)$ need not be decomposed further: we already know that $\Box(P \vee Q)$ is true at u.

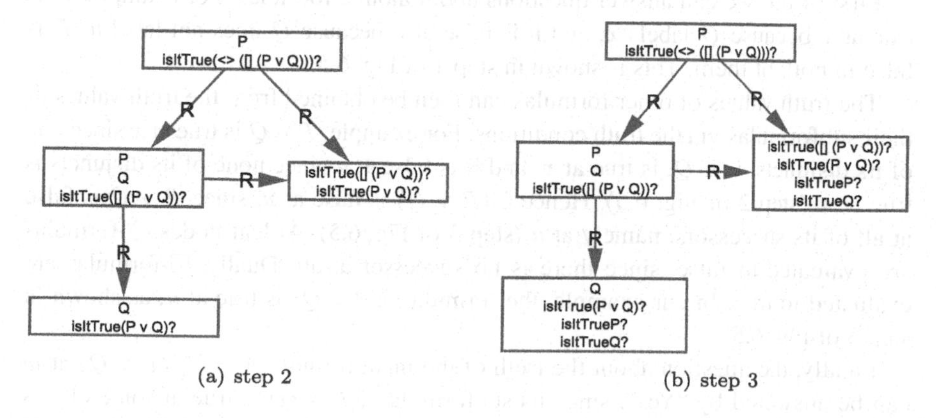

(a) step 2 (b) step 3

Fig. 6.4 Top-down questions phase, steps 2 and 3

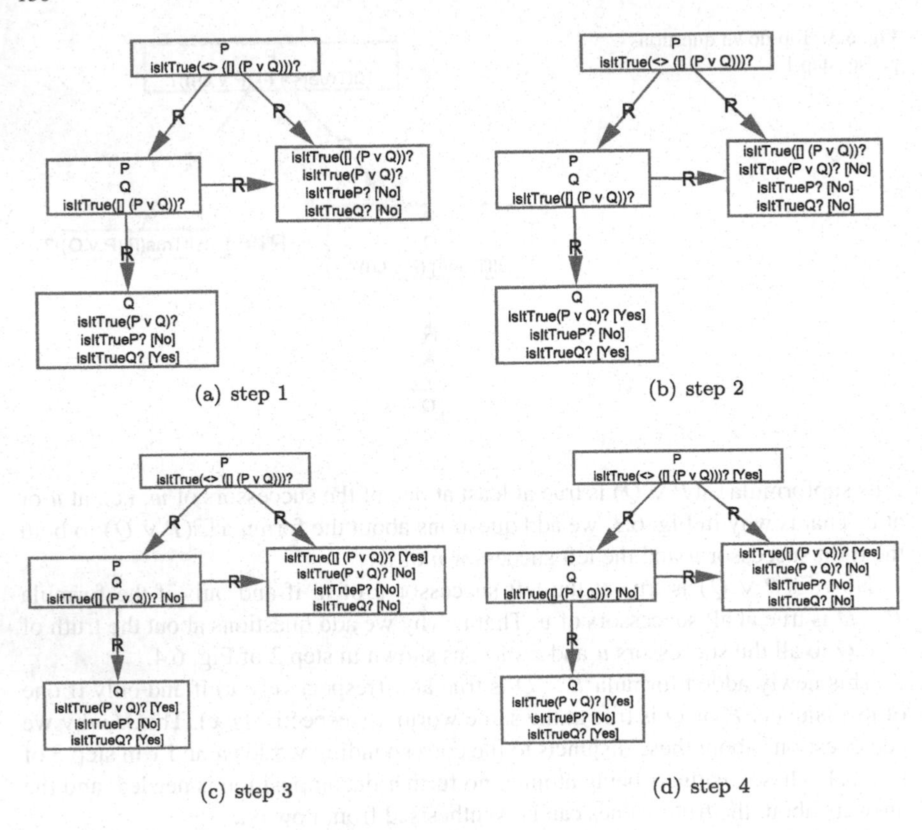

(a) step 1 (b) step 2

(c) step 3 (d) step 4

Fig. 6.5 Bottom-up answer phase

Bottom-Up Answers Phase We are now ready to answer the questions about the truth of the subformulas, which will be done by means of backtracking.

First of all, we can answer questions about atomic formulas. For example, Q is true at x because Q labels x, and it is false at u because Q does not label u. P is false in both of them. This is shown in step 1 of Fig. 6.5.

The truth values of other formulas can then be obtained from the truth values of their subformulas via the truth conditions. For example, $P \vee Q$ is true at x since one of its disjuncts, viz. Q, is true at x, and it is false at u since none of its disjuncts is true at u (step 2 of Fig. 6.5). Hence $\Box(P \vee Q)$ is false at v, since $P \vee Q$ is false at all of its successors, namely at u (step 3 of Fig. 6.5). At leaf nodes, $\Diamond$-formulas are evaluated to false, since there is no successor at all. Dually, $\Box$-formulas are evaluated to true. In our example, the formula $\Box(P \vee Q)$ is true at u, as shown in step 3 of Fig. 6.5.

Finally, the question about the truth of the input formula $A = \Diamond\Box(P \vee Q)$ at w can be answered by "Yes", since its subformula $\Box(P \vee Q)$ is true at some of w's successors, namely u (step 4 of Fig. 6.5).

6.2 How to Implement Model Checking in LoTREC

Let us implement the above model checking procedure in LoTREC.

Defining a Model Checking Problem We start by building a model in the same way as we already did in Sect. 1.7. We then insert a question about the formula to be checked into the root node.

The construction of a model consists in creating a finite graph. In LoTREC, this can be done in three different ways:

1. using a rule,
2. loading a previously saved model file (via the Premodels menu), or
3. adding nodes, edges and formulas using the graphical user interface.

We have already shown in Sect. 1.7 how to build a model by means of a rule. The following rule builds the model of Fig. 6.1.

```
Rule BuildExampleGraph
  isNewNode w

  createNewNode u
  createNewNode v
  createNewNode x
  link w u R
  link w v R
  link v u R
  link v x R
  add w P
  add v P
  add v Q
  add x Q
```

This will be the first rule of the model checking strategy.

Once the model is built, we enter the input formula A by adding the question concerning the formula A to the root world w. We may do so within the rule building the model (or after loading the model from a file or drawing it by hand). We should distinguish questions about formulas from the atoms labelling the nodes of the model; we therefore prefix the questions by a special unary connective. The connective identifying questions is defined as follows:

Name	Arity	Display
isItTrue	1	isItTrue(_)?

In order to add the formula `<> [] (P v Q)` to the model of Fig. 6.1 we insert the action add w isItTrue pos nec or P Q somewhere in the rule BuildExampleGraph (the position does not matter). If we run LoTREC with a simple strategy applying this rule just once we obtain the model of Fig. 6.2.

Top-Down Rules The top-down steps presented in Sect. 6.1 can be achieved automatically by an appropriate set of rules transforming the question about the truth value of the input formula into smaller questions about the truth values of its subformulas.

For example, to check the truth of a formula of the form $\neg A$ at world w, the truth of A should be checked at w; and to check formulas of the form $A \wedge B$ or of the form $A \vee B$, both formulas A and B should be checked. The following rules do this analysis:

```
Rule Not_TopDown
  hasElement w isItTrue not variable A

  add w isItTrue variable A
```

```
Rule And_TopDown
  hasElement w isItTrue and variable A variable B

  add w isItTrue variable A
  add w isItTrue variable B
```

```
Rule Or_TopDown
  hasElement w isItTrue or variable A variable B

  add w isItTrue variable A
  add w isItTrue variable B
```

Rules dealing with formulas obtained with other boolean connectives such as $\rightarrow$ and $\leftrightarrow$ can be defined in the same way.

As for modal formulas, we check their subformulas in successor nodes as follows:

```
Rule Nec_TopDown
  hasElement w isItTrue nec variable A
  isLinked w u R

  add u isItTrue variable A
```

```
Rule Pos_TopDown
  hasElement w isItTrue pos variable A
  isLinked w u R

  add u isItTrue variable A
```

We have to call the above rules repeatedly in oder to entirely decompose the input formula into its constituents. This is what the following strategy does:

```
Strategy Top_Down_Strategy
  repeat
    Not_TopDown
    And_TopDown
    Or_TopDown
    Nec_TopDown
    Pos_TopDown
  end
end
```

Bottom-Up Rules Now that we have analysed the input formula we are ready to evaluate the truth values of its subformulas. We signal the truth value of A by marking `isItTrue(A)?` by `Yes` to signal that it is true, and by `No` otherwise. The first answers are for atoms and are obtained by the following rules:

```
Rule Atom_True_BottomUp
  hasElement w isItTrue variable A
  isAtomic variable A
  hasElement w variable A

  markExpressions w isItTrue variable A Yes
```

```
Rule Atom_False_BottomUp
  hasElement w isItTrue variable A
  isAtomic variable A
  hasNotElement w variable A

  markExpressions w isItTrue variable A No
```

Once we have obtained the answers for the atomic formulas, we can synthesise the answers for complex formulas just as in the bottom-up steps of Sect. 6.1. For negated formulas we write:

```
Rule Not_True_BottomUp
  hasElement w isItTrue not variable A
  isMarkedExpression w isItTrue variable A No

  markExpressions w isItTrue not variable A Yes
```

```
Rule Not_False_BottomUp
  hasElement w isItTrue not variable A
  isMarkedExpression w isItTrue variable A Yes

  markExpressions w isItTrue not variable A No
```

For conjunctions:

```
Rule And_True_BottomUp
  hasElement w isItTrue and variable A variable B
  isMarkedExpression w isItTrue variable A Yes
  isMarkedExpression w isItTrue variable B Yes

  markExpressions w isItTrue and variable A variable B Yes
```

```
Rule And_FalseLeft_BottomUp
  hasElement w isItTrue and variable A variable B
  isMarkedExpression w isItTrue variable A No

  markExpressions w isItTrue and variable A variable B No
```

```
Rule And_FalseRight_BottomUp
  hasElement w isItTrue and variable A variable B
  isMarkedExpression w isItTrue variable B No

  markExpressions w isItTrue and variable A variable B No
```

For disjunctions:

```
Rule Or_False_BottomUp
  hasElement w isItTrue or variable A variable B
  isMarkedExpression w isItTrue variable A No
  isMarkedExpression w isItTrue variable B No

  markExpressions w isItTrue or variable A variable B No
```

```
Rule Or_TrueLeft_BottomUp
  hasElement w isItTrue or variable A variable B
  isMarkedExpression w isItTrue variable A Yes

  markExpressions w isItTrue or variable A variable B Yes
```

```
Rule Or_TrueRight_BottomUp
  hasElement w isItTrue or variable A variable B
  isMarkedExpression w isItTrue variable B Yes

  markExpressions w isItTrue or variable A variable B Yes
```

As to modal formulas of the form `[] A`, we have to check whether `A` is marked by `Yes` in all successors. If it is the case, then the question `isItTrue([] A)` can be tagged by Yes. Observe that isItTrue([] A) has also to be tagged by Yes when there are no successors at all. The other way round, in order to tag isItTrue([] A) by No it is sufficient that A is tagged by No in some of w's successors. In order to deal with the first case we need a new LoTREC keyword:

```
isMarkedExpressionInAllChildren w A R M
```

checks whether every R-successor of node w contains some label A that is marked by M.

```
Rule Nec_True_BottomUp
  hasElement w isItTrue nec variable A
  isMarkedExpressionInAllChildren w isItTrue variable A R Yes

  markExpressions w isItTrue nec variable A Yes


Rule Nec_False_BottomUp
  hasElement w isItTrue nec variable A
  isLinked w u R
  isMarkedExpression u isItTrue variable A No

  markExpressions w isItTrue nec variable A No
```

Dually, ◊-formulas are evaluated by means of the following two rules:

```
Rule Pos_False_BottomUp
  hasElement w isItTrue pos variable A
  isMarkedExpressionInAllChildren w isItTrue variable A R No

  markExpressions w isItTrue pos variable A No


Rule Pos_True_BottomUp
  hasElement w isItTrue pos variable A
  isLinked w u R
  isMarkedExpression u isItTrue variable A Yes

  markExpressions w isItTrue pos variable A Yes
```

Let us put the set of bottom-up rules in a strategy:

```
Strategy Bottom_Up_Strategy
  repeat
    Atom_True_BottomUp
    Atom_False_BottomUp
    Not_True_BottomUp
    Not_False_BottomUp
    And_True_BottomUp
```

```
    And_FalseLeft_BottomUp
    And_FalseRight_BottomUp
    Or_False_BottomUp
    Or_TrueLeft_BottomUp
    Or_TrueRight_BottomUp
    Nec_True_BottomUp
    Nec_False_BottomUp
    Pos_False_BottomUp
    Pos_True_BottomUp
  end
end
```

The Main Strategy It remains to integrate the rule building the model and inserting the input formula with the top-down and the bottom-up phases. This is achieved by the following simple strategy that calls `BuildExampleGraph`, `Top_Down_Strategy`, and `Bottom_Up_Strategy` in sequence.

```
Strategy Model_Checking_Strategy
  BuildExampleGraph
  Top_Down_Strategy
  Bottom_Up_Strategy
end
```

The `BuildExampleGraph` can be omitted from the strategy if the model is built without using the rule, e.g. by using the graphical user interface.

Exercise 78 Instead of using the auxiliary connective `isItTrue` one may as well use marks in order to identify the questions. Modify the rules accordingly, using the LoTREC action `mark` and the LoTREC condition `isMarked`.

6.3 Conclusion

In this chapter, we have given a LoTREC procedure to solve the model checking problem. It is suitable for all the monomodal logics that we have seen up to now. In order to deal with multimodal versions of these logics, our procedure should be extended. The extension is simple and proceeds just as we did in Chap. 3 to extend the model construction method for the monomodal **K** to the multimodal $\mathbf{K}_n$: in the top-down and bottom-up rules, we prefix each `variable R` in the appropriate places: we replace all occurrences of `nec variable A`, `pos variable A`, and `R` by `nec variable R variable A`, `pos variable R variable A`, and `variable R`, respectively.

Dealing with formulas from other logics, having other sets of special connectives or having different semantics may require more effort to adapt the above method.

In the next chapter, the above model checking strategy will be inserted into a procedure for model construction.

Chapter 7
Modal Logics with Transitive Closure

Chapter 5 was about several logics of frames whose accessibility relations are transitive. The present chapter is about logics of frames having two accessibility relations one of which is the transitive closure of the other. The closure will also be reflexive for all the logics of the present chapter (except in Exercise 84). The usual symbol for reflexive and transitive closure being '*', we may write these two relations $R(I)$ and $R(I)^*$, for some I. The language of these logics has a modal operator $[I]$ that is interpreted by $R(I)$ and another modal operator $[I^*]$ that is interpreted by $R(I)^*$.

The reflexive and transitive closure of $R(I)$ is the smallest relation that is reflexive and transitive and contains $R(I)$. It can also by defined more constructively:

$$R(I)^* = \{ \langle w, u \rangle \mid \text{there is a finite sequence of worlds } \langle v_0, \ldots, v_n \rangle \text{ such that } v_0 = w,\ v_n = u, \text{ and } \langle v_i, v_{i+1} \rangle \in R(I) \text{ for } 0 \leq i < n \}$$

Transitive closure is important for example in the logical analysis of common knowledge. We say that common knowledge between Ann and Bob that A is the case when the following infinite sequence of statements is the case:

- Ann knows that A;
- Bob knows that A;
- Ann knows that Bob knows that A;
- Bob knows that Ann knows that A;
- Ann knows that Bob knows that Ann knows that A;
- and so on *ad infinitum*.

In the semantics, "common knowledge of Ann and Bob that A" means that A is true at all worlds that can be reached from the actual world by any finite sequence of $R(\mathsf{Ann})$ edges and $R(\mathsf{Bob})$ edges. In other words, A holds in all worlds reachable from the actual world by the reflexive and transitive closure of $R(\mathsf{Ann}) \cup R(\mathsf{Bob})$.

Another application is software verification. Let I be a piece of code of a computer program written in some programming language. As we have seen in Sect. 1.1, the behavior of a single execution of I can be modelled by the relation $R(I)$ between possible worlds (that are viewed as program states). We also want to be able

O. Gasquet et al., *Kripke's Worlds*, Studies in Universal Logic,
DOI 10.1007/978-3-7643-8504-0_7, © Springer Basel AG 2014

to model the iteration of such a piece of code inside a *while* or *repeat* loop. This can be done in a natural way by means of the reflexive and transitive closure $R(I)^*$ of $R(I)$. When we reason about the properties of *while* or *repeat* loops we have to be able to talk about that reflexive and transitive closure in the logical language. There are basically two families of modal logics allowing us to do that. The first are temporal logics, most prominently linear-time temporal logic **LTL**, which is widely used for software verification [BK08]. **LTL** formulas do not talk about the programs themselves, but only about the temporal properties that can be associated to them. The language contains formulas such as $\mathsf{X}A$, read "Next A" and $\mathsf{G}A$, read "Globally A" or "Henceforth A"; The accessibility relations for these two modal operators naturally involve reflexive and transitive closure: the relation for G is the reflexive and transitive closure of the relation for X.

The second family of logics are dynamic logics [HKT00], whose basic logic is propositional dynamic logic **PDL**. The languages of these logics offer constructions such as $[I]A$ where I is a program and A is a property, read "After all executions of the program I, the property A holds." The program I may be complex and may in particular be built with the iteration operator $*$. The latter is also called the *Kleene star*.

The tableaux procedures for logics with transitive closure of the present chapters are the most complex in this book and require us to combine techniques from the previous chapters: we put together the standard tableau rules of Chaps. 3 and 4, the blocking technique of Chap. 5, and the model checking procedure of Chap. 6.

We address model construction for **LTL** formulas in Sect. 7.1. In Sect. 7.2 we address the model construction for propositional dynamic logic **PDL**; in the end of that section we put the tableaux procedure for the logic of common knowledge in terms of an exercise.

7.1 LTL: Linear-Time Temporal Logic

When we use Kripke models to reason about temporal properties then we view possible worlds as moments in time and the temporal accessibility relation provides an order on the set of moments. A moment u which is accessible from a moment w is viewed as being in the future of w. The precise nature of the accessibility relation depends on our view of time.

- Time might be linear or branching. If time is linear then the future is completely determined. On the contrary, when time is branching then there can be several future courses of time. For instance, suppose Ann is in love with Bob. When the model of time is linear then they will either marry in the future or not, but only one of these courses of time is possible. Branching time allows for one possible future where Ann will marry Bob and another possible future where Ann will not marry Bob. From that perspective, time should be branching in the future. However, at least when one considers deterministic programs one may safely suppose that the future is linear.

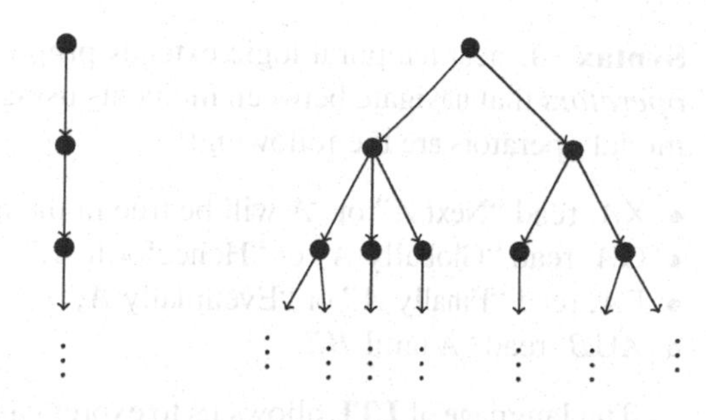

Fig. 7.1 Linear time (*left*) and branching time (*right*) for the case of discrete time

- Time may be discrete or not. When time is discrete then we can talk about the successors of a given moment w (and if time is moreover linear then there is a unique successor). When time is dense then between two distinct moments there is always at least one moment. Dense time may be continuous or not: when it is continuous it has (locally) the same topology as the line of real numbers.
- Time might or not have endpoints in the future of in the past.

Figure 7.1 illustrates the difference between linear and branching time for the discrete case.

In linear-time temporal logic, abbreviated **LTL**, time is considered to be linear, discrete and without endpoints in the future. (It may or may not have endpoints in the past.) **LTL** is used a lot in program verification. The most prominent logic of branching and discrete time is Computational Tree Logic **CTL**.

7.1.1 Syntax and Semantics

Let us define the models of linear-time temporal logic **LTL** and its syntax.

Models The models of **LTL** have an accessibility relation that is serial and deterministic. So **LTL** models are of the form $M = (W, R, V)$ where R is a unique relation that is serial and deterministic. The relation R is associated to the 'next' operator X, while its reflexive and transitive closure R^* (which we omit from the model description by convention) is associated to the 'global' operator G.

Due to seriality and determinism of R, for every world w there exists a unique world w' such that wRw'. We call w' the successor of w, or the *next* moment after w. R can be actually viewed as a function that provides a linear discrete timeline that is isomorphic to the set of natural numbers $\mathbb{N}$. For a given moment w we may write the *path*

$$\langle w_0, w_1, w_2, \dots \rangle$$

in order to designate the sequence of moments that are in the future of w, where $w_0 = w$, w_1 is the successor of w_0, w_2 is the successor of w_1, etc.

Syntax Linear temporal logic extends propositional logic with a set of *temporal operators* that navigate between moments using the accessibility relation. The basic modal operators are the following:[1]

- $\mathsf{X}A$, read "Next A" or "A will be true in the next state;"
- $\mathsf{G}A$, read "Globally A" or "Henceforth A;"
- $\mathsf{F}A$, read "Finally A" or "Eventually A;"
- $A\mathsf{U}B$, read "A until B."

The language of **LTL** allows us to express a bunch of properties that are interesting for program verification:

- *Safety*: "Something bad will never happen," e.g.

$$\mathsf{G}\neg(\mathsf{KeyInCar} \wedge \mathsf{CarClosed} \wedge \neg\mathsf{DriverInCar})$$

- *Liveness*: "Something good will happen," e.g.

$$\mathsf{G}(\mathsf{PhdStarted} \rightarrow \mathsf{F}\,\mathsf{PhdGranted})$$

- *Fairness*: "If something is attempted/requested infinitely often, then it will be successful/allocated infinitely often," e.g.

$$(\mathsf{GF}\ \mathsf{ClientAsksForPrinter}) \rightarrow (\mathsf{GF}\ \mathsf{ServerAllocatesPrinter})$$

Semantics Let $M = (W, R, V)$ be a **LTL** model and let $w \in W$. The truth conditions for the modal operators of **LTL** are as follows.[2]

$$\begin{array}{ll} M, w \Vdash \mathsf{X}A & \text{iff for every } u, \text{ if } (w,u) \in R \text{ then } M, u \Vdash A; \\ M, w \Vdash \mathsf{G}A & \text{iff for every } u, \text{ if } (w,u) \in R^* \text{ then } M, u \Vdash A; \\ M, w \Vdash \mathsf{F}A & \text{iff there is a } u \text{ such that } (w,u) \in R^* \text{ and } M, u \Vdash A; \end{array}$$

---

[1] There exist alternative notations for the first three operators: $\bigcirc A$ instead of $\mathsf{X}A$, $\Box A$ instead of $\mathsf{F}A$, and $\Diamond A$ instead of $\mathsf{G}A$.

[2] A more standard formulation uses paths. Let $\langle w_0, w_1, \ldots\rangle$ be the sequence of moments that are in the future of $w_0 = w$.

$$\begin{array}{rl} M, w_0 \Vdash \mathsf{X}A & \text{iff } M, w_1 \Vdash A; \\ M, w_0 \Vdash \mathsf{F}A & \text{iff there is a natural number } i \text{ such that } M, w_i \Vdash A; \\ M, w_0 \Vdash \mathsf{G}A & \text{iff for all natural numbers } i \text{ we have } M, w_i \Vdash A; \\ M, w_0 \Vdash A\mathsf{U}B & \text{iff there is a natural number } i \text{ such that } M, w_i \Vdash B \text{ and} \\ & \text{for all } 0 \le j < i \text{ we have } M, w_j \Vdash A. \end{array}$$

$M, w \Vdash A\mathsf{U}B$ iff there is a v such that $M, v \Vdash B$ and for all u such that $(w, u) \in R^*$ and $(u, v) \in R^*$ we have $M, u \Vdash A$.

The next series of equivalences will be used by the tableaux method for **LTL**.

$$\neg \mathsf{X}A \leftrightarrow \mathsf{X}\neg A$$

$$\neg \mathsf{G}A \leftrightarrow \mathsf{F}\neg A$$

$$\neg \mathsf{F}A \leftrightarrow \mathsf{G}\neg A$$

$$\mathsf{G}A \leftrightarrow A \wedge \mathsf{XG}A$$

$$\mathsf{F}A \leftrightarrow A \vee \mathsf{XF}A$$

$$A\mathsf{U}B \leftrightarrow B \vee \big(A \wedge \mathsf{X}(A\mathsf{U}B)\big)$$

$$\top\mathsf{U}A \leftrightarrow \mathsf{F}A$$

$$\neg(A\mathsf{U}B) \leftrightarrow (\neg B \wedge \neg A) \vee \big(\neg B \wedge \mathsf{X}\neg(A\mathsf{U}B)\big)$$

As the reader may check, the following two equivalences are valid in **LTL**:

$$\mathsf{G}A \leftrightarrow \top\mathsf{U}A$$

$$\mathsf{F}A \leftrightarrow \neg\mathsf{G}\neg A$$

$$\leftrightarrow \neg(\top\mathsf{U}\neg A)$$

They allow consideration of G and F as abbreviations and present **LTL** in terms of a language whose only modal operators are X and U. We have however kept these operators in the subsequent presentation of the tableaux method in order to improve its understanding.

Before we start, let us also remark that Exercise 47 in Sect. 4.5 is about the logic of a single serial and deterministic accessibility relation. The **LTL** relation actually has these two properties, and the tableaux procedure of that exercise also provides a procedure for the fragment of **LTL** formulas whose only modal operator is X.

7.1.2 Model Construction for LTL in LoTREC: Saturating Premodels

The **LTL** connectives can be defined in `LoTREC` as explained in Sect. 2.2. The boolean connectives have the same truth conditions as in propositional logic, and the rules dealing with them are the same as those for the modal logics we have seen up to now. Remember that they are grouped in the substrategy `BooleanRules`. In the rest of the section we focus on the rules for the temporal operators.

Rules for Negated Formulas To save space we deal with negations of modalities in a way different from what we did up to now and add the equivalent formula where negation is pushed inside. For example, when a node contains `~X A` then we add `X ~A`, and when it contains `~[I] A` we add `<I> ~A`. Here is the corresponding list of rules, one per modal operator.

```
Rule NotNext_rewrite
  hasElement w not next variable A

  add w next not variable A


Rule NotFinally_rewrite
  hasElement w not finally variable A

  add w globally not variable A


Rule NotGlobally_rewrite
  hasElement w not globally variable A

  add w finally not variable A


Rule NotUntil_rewrite
  hasElement w not until variable A variable B

  add w or and not variable B not variable A
           and not variable B
               next not until variable A variable B
```

The rule for the 'until' operator corresponds to one of the equivalences listed above

$$\neg(A \mathsf{U} B) \leftrightarrow (\neg B \land \neg A) \lor \left(\neg B \land \mathsf{X}\neg(A \mathsf{U} B)\right).$$

Note that the formulas that we add in this way are not subformulas modulo negation of the original formula. The latter definition can however be extended in order to include such subformulas, yielding the Fisher-Ladner closure of the formula.

Rule for X The rule dealing with the 'next' operator X is similar to the way we proceeded for $\mathbf{K.Alt}_1$ in Sect. 4.4: given a world w, we create a unique successor of w, say u, and link it to w by the accessibility relation R. We have:

```
Rule Next_deterministic1
  hasElement w next variable A
  hasNoSuccessor w R

  createNewNode u
```

```
    link w u R

Rule Next_deterministic2
    hasElement w next variable A
    isLinked w u R

    add u variable A
```

As we have explained in Sect. 4.4, rules creating a single successor (there: Pos_deterministic1 and NotNec_deterministic1) should be prefixed by the applyOnce keyword in the strategy. The rule Next_deterministic2 appropriately propagates the subformula A of formulas of the form X A to the unique successor.

Rule for G According to the truth conditions, $\mathsf{G}A$ is equivalent to $A \wedge \mathsf{XG}A$. Hence we define a rule which, for every formula G A found at node w, adds both A and X G A to w. In a sense, this rule checks A at w and postpones the check of G A from w to the successor of w.

```
Rule Globally
    hasElement w globally variable A

    add w variable A
    add w next globally variable A
```

Note that the rule Globally is not analytic modulo negation. The formula that is added is however in the so-called *Fisher-Ladner closure* of the original formula. This will ensure termination.

Eventualities and Their Fulfilment We pause here and introduce the concept of an *eventuality*. These are formulas whose truth condition requires the existence of some world in the future where A is true. The eventualities of our temporal language are the formulas of the form $\mathsf{F}A$ and $B\mathsf{U}A$.

We say that an eventuality F A or B U A is *fulfilled* at a node w of a premodel M if and only if the formula A labels either w or one of the R-descendant nodes of w in M. If w is a loop node included in an ancestor node u then w is going to be identified with u; therefore if F A is fulfilled at u it is also fulfilled at w. The definition for B U A is similar.

Clearly, when a tableaux procedure produces an open saturated premodel then all F A and all A U B should be fulfilled in order to turn the premodel into a model. In K_Strategy, the Pos rule guarantees immediate fulfilment of every eventuality <> A in one of the successors. In K4_Strategy, the Pos_ifUnblocked rule guarantees either immediate fulfilment in one of the successors of w, or 'indirect fulfilment' in the successor node of some ancestor of w containing w (to which w will be identified when the premodel is transformed into a **K4** model). In contrast the **LTL** rules for the eventualities F and U that we shall see now do not guarantee fulfilment.

Rule for F We handle the F operator in a way that is similar to the G operator: from the equivalence $\mathsf{F}A \leftrightarrow (A \vee \mathsf{XF}A)$ we obtain the following rule.

```
Rule Finally
  hasElement w finally variable A

  duplicate premodel_copy
  add w variable A
  add premodel_copy.w next finally variable A
```

So we copy the actual premodel and call it `premodel_copy`, just as we do for disjunctions. We then add `A` to the node `w` of the current premodel (thus directly fulfilling `F A` of `w`) and we add `X F A` to the world `w` of `premodel_copy`. When the `Next_deterministic2` rule will be applied at a later stage then the formula `F A` is added to the successor of `w` in `premodel_copy` (if there is any). In this way, the fulfilment of `F A` in `premodel_copy` is *postponed* to the next step of the model construction process.

Rule for U The 'until' connective U is dealt with by the following rule:

```
Rule Until
  hasElement w until variable B variable A

  duplicate premodel_copy
  add w variable A
  add premodel_copy.w variable B
  add premodel_copy.w next until variable B variable A
```

Just as the rule `Finally` does, this rule makes a copy of the actual premodel and calls it `premodel_copy`. It adds `A` to the node `w` of the current premodel, thereby directly fulfilling `B U A`. In the world `w` of `premodel_copy`, it adds instead `B` and `X (B U A)`. The latter will at a later stage trigger the `Next_deterministic2` rule, which will add `B U A` to the successor of `w` in `premodel_copy` (if there is any). In this way, the fulfilment of the formula `B U A`—viz. by a descendant node with the formula `A`, and `B`-nodes in between—is postponed in `premodel_copy` to the next step of the model construction process.

7.1.3 LoTREC Rules for LTL: Termination by Node-Inclusion Tests

Let us put the sequence of the above tableau rules inside a `repeat` loop, in any order. The resulting strategy faces the same problem as our naive strategy for logic **K4** in Sect. 5.1: there exist formulas for which the strategy does not terminate.

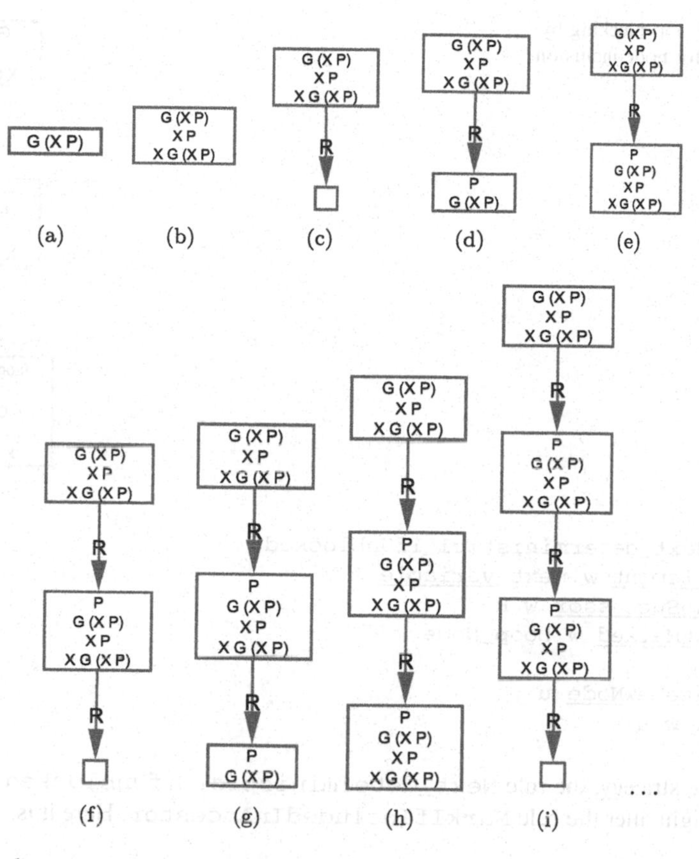

Fig. 7.2 No loop test: an example of an infinite model construction

Figure 7.2 contains an example. It starts with the formula `G X P` (premodel (a)). This yields a premodel with `P` and the same original formula `G X P` (premodel (d)). The latter leads to a new successor with `P` and the same initial formula `G X P` (premodel (g)), and so on.

To avoid such loops we add to the above naive strategy a rule checking for inclusion in some ancestor:

```
Rule MarkIfIncludedInAncestor
  isNewNode u
  isAncestor w u
  contains w u

  mark u Loop_Node
```

and we block every node that is included in some ancestor node by changing the above rule `Next_deterministic1` as follows:

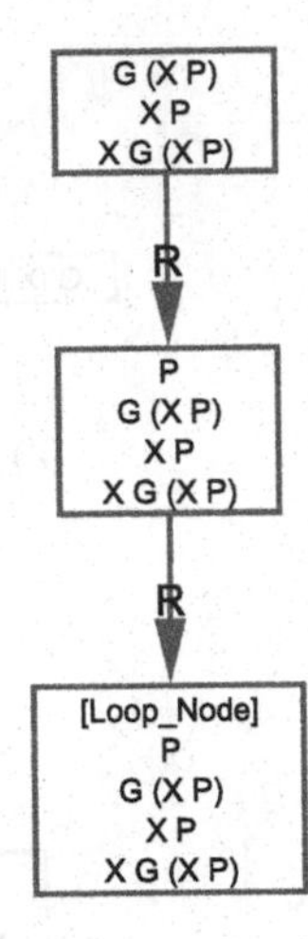

Fig. 7.3 Loop-blocking by checking for node-inclusion

```
Rule Next_deterministic1_ifUnblocked
  hasElement w next variable A
  hasNoSuccessor w R
  isNotMarked w Loop_Node

  createNewNode u
  link w u R
```

In the strategy, the rule `Next_deterministic1_ifUnblocked` should be called right after the rule `MarkIfIncludedInAncestor`. Here it is:

```
Strategy LTL_Saturation_Strategy
  repeat
    repeat
      BooleanRules
      NotNext_rewrite
      NotGlobally_rewrite
      NotFinally_rewrite
      NotUntil_rewrite
      Globally
      Finally
      Until
    end
    MarkIfIncludedInAncestor
    applyOnce Next_deterministic1_ifUnblocked
    Next_deterministic2
  end
end
```

Running `LTL_Saturation_Strategy` on the above formula G X P terminates, as shown in Fig. 7.3.

Exercise 79 Run the naive strategy on the formula FP (use the step-by-step mode). Then run `LTL_Saturation_Strategy`.

Remark 12 Note that we may change all the other rules by checking whether the node is not blocked. They then become inapplicable on nodes that are tagged as `Loop_Node`. Such nodes are then completely blocked. This is particularly interesting for the duplicating rules `Or`, `Finally`, and `Until`. Note that the rule `Mark-IfIncludedInAncestor` should only be called *after* these rules: otherwise we may tag nodes too early, i.e., before creating new successors.

The above blocking technique is the same as the one we used for **K4** in Sect. 5.1 in order to guarantee the termination of the model construction method for **K4**. It also guarantees termination in the present case of **LTL**. However, the resulting tableaux procedure is still not suitable, as we will see later on. But let us address another problem first: *some eventualities may be not fulfilled.*

7.1.4 LoTREC Rules for LTL: Fulfilment of Eventualities

In this subsection we deal with a problem that comes with the blocking technique: there are open premodels to which the `Finally` or `Until` rule cannot be applied because the nodes under concern are blocked. These premodels are therefore saturated. However, they cannot be transformed into a model because one of these eventualities is not fulfilled. Here is an example.

Inextensible Premodels Due to Unfulfilled Eventualities The formula G$\neg P \wedge$ FP is unsatisfiable, and a correct tableaux procedure should only return closed premodels for this formula. However, Fig. 7.4 shows that one of the saturated premodels that is returned by `LTL_Saturation_Strategy` is open, viz. premodel.2.2. This open premodel is blocked since a loop is detected after checking for node inclusion. The formula `F P` is not fulfilled in this open premodel since `P` is false in any world of this premodel. Hence this premodel cannot be transformed into a model of the input formula. We call it therefore an *inextensible* premodel.

We conclude that in **LTL** we have to perform a further check on open saturated premodels: we have to verify that all the eventualities `F A` and `A U B` are fulfilled.

Checking the Fulfilment of Eventualities In order to easily access the loop ancestor we change the `MarkIfIncludedInAncestor` rule as follows:

```
Rule MarkAndLink_ifIncludedInAncestor
  isNewNode u
  isAncestor w u
  contains w u

  mark u Loop_Node
  link u w Loop
```

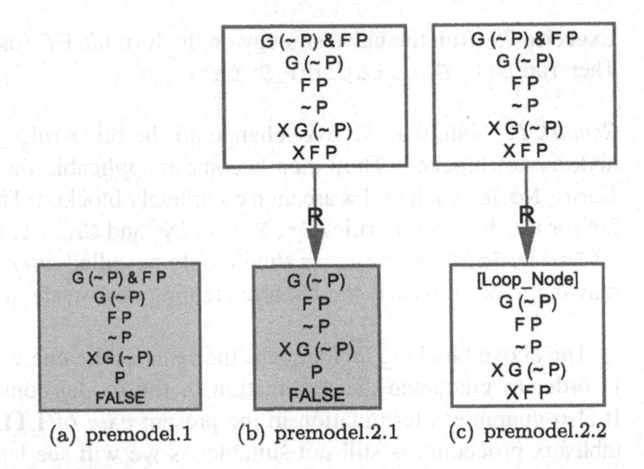

Fig. 7.4 An example of an open (blocked) premodel for an unsatisfiable formula

The rule links nodes to ancestor nodes containing it by a special edge that is labelled `Loop`. Such edges can be avoided, but are often convenient to 'read' the premodels that are returned by `LoTREC`.

We then check the fulfilment of eventualities using the model-checking technique of Chap. 6. In order to distinguish fulfilled from unfulfilled '`F`' eventualities we repeatedly call the following rules:

```
Rule Finally_fulfilledAtSameNode
  hasElement w finally variable A
  hasElement w variable A

  markExpressions w finally variable A Fulfilled


Rule Finally_fulfilledAtDescendant
  hasElement w finally variable A
  isAncestor w u
  hasElement u variable A

  markExpressions w finally variable A Fulfilled


Rule Finally_fulfilledAtLoopAncestor
  hasElement u finally variable A
  isLinked u w Loop
  isMarkedExpression w finally variable A Fulfilled

  markExpressions u finally variable A Fulfilled
```

Exercise 80 Define rules marking fulfilled '`U` eventualities.' Call them

`Until_fulfilledAtSameWorld`

```
Until_fulfilledAtDescendant
Until_fulfilledAtLoopAncestor
```

The rules marking fulfilled eventualities are put at the end of the **LTL** strategy, after the saturating repeat loop. Once they have been applied, unfulfilled eventualities will be left without being tagged Fulfilled. In the next step we are going to identify nodes containing an unfulfilled eventuality by marking them by Inextensible_Premodel. This is done as follows:

```
Rule Finally_unfulfilledIsInextensible
  hasElement w finally variable A
  isNotMarkedExpression w finally variable A Fulfilled

  mark w Inextensible_Premodel
```

```
Rule Until_unfulfilledIsInextensible
  hasElement w until variable A variable B
  isNotMarkedExpression w until variable A variable B
    Fulfilled

  mark w Inextensible_Premodel
```

Again, one can do without such tags, but they are often useful in order to 'read' a premodel that is returned by LoTREC.

The following two rules propagate the Inextensible_Premodel tag to every node of a given inextensible premodel, including the node containing the input formula. Again, adding this information is not mandatory, but it allows the user to identify premodels with unfulfilled eventualities.

```
Rule PropagateInextensibilityUp
  isMarked w Inextensible_Premodel
  isLinked u w R

  mark u Inextensible_Premodel
```

```
Rule PropagateInextensibilityDown
  isMarked w Inextensible_Premodel
  isLinked w u R

  mark u Inextensible_Premodel
```

Let us add the above rules at the end of LTL_Saturation_Strategy and run it on the same example formula $\mathsf{G}\neg P \wedge \mathsf{F}P$. LoTREC returns the same two closed premodels (premodel.1 and premodel.2.1) of Fig. 7.4, whereas the third open premodel (premodel.2.2) is reported as being inextensible, as shown in Fig. 7.5.

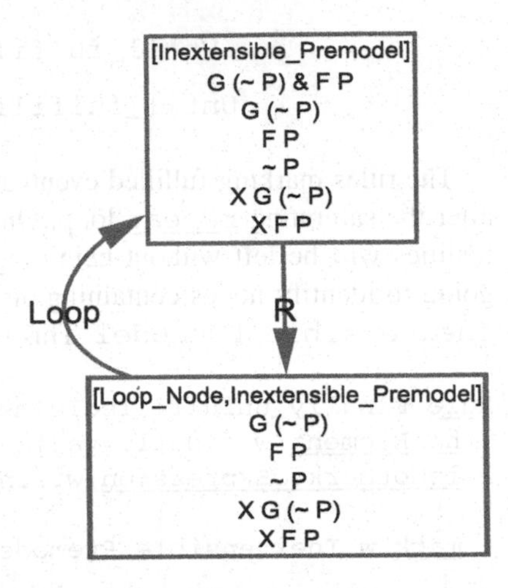

Fig. 7.5 The only open premodel for the formula $\mathsf{G}\neg P \wedge \mathsf{F}P$ is reported to be inextensible since the eventuality $\mathsf{F}P$ is not fulfilled (in both nodes)

7.1.5 LoTREC Rules for LTL: Termination by Node-Equality Test

There is another specificity of **LTL** loop checking in comparison with **K4**. Checking for node inclusion is sufficient in the case of **K4**, as we have seen in Sect. 5.1, while it is no longer suitable in the case of **LTL**. Let us show this by means of an example.

Consider the formula $\mathsf{F}P \wedge \neg P \wedge \mathsf{X}\neg P$. Figure 7.6 shows the result of running our method (with node-inclusion loop test and eventuality fulfilment check). It yields three premodels. Two of them are closed (premodel.1 and premodel.2.1), while the third (premodel.2.2) is open but blocked due to node-inclusion and then marked as an inextensible because the eventuality `F P` is not fulfilled.

Nevertheless, this formula is satisfiable, as shown in Fig. 7.7. There, the premodel could have been obtained by our method if we had continued developing premodel.2.2—that is blocked by node-inclusion—just one further step.

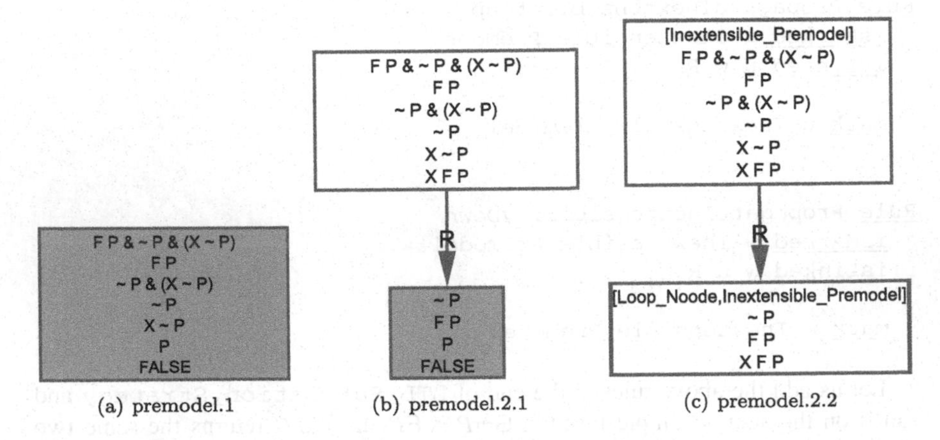

Fig. 7.6 Blocking a node included in an ancestor node guarantees termination, but does not allow one to find an open saturated premodel for $\mathsf{F}P \wedge \neg P \wedge \mathsf{X}\neg P$

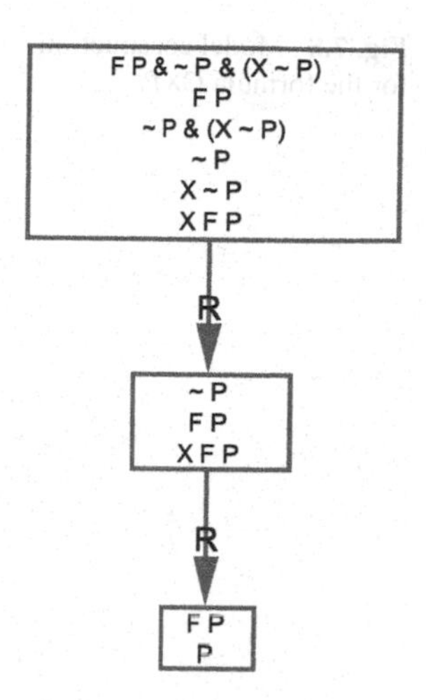

Fig. 7.7 The premodel of the formula $\mathsf{F}P \wedge \neg P \wedge \mathsf{X}\neg P$, which is missed by the method with node inclusion test, and which could have been obtained by developing furthermore the premodel 2.2 of Fig. 7.6

Node inclusion is not strong enough to stop the development of a node: the ancestor node should contain exactly the same constraints as the node we would like to block. We therefore should identify loops by checking for *node equality*. This is what the following rule does:

```
Rule Mark_ifEqualToAncestor
 isNewNode u
 isAncestor w u
 contains w u
 contains u w

 mark u Loop_Node
```

This rule replaces `MarkAndLink_ifIncludedInAncestor` in `LTL_Saturation_Strategy`. We may also add to this rule the action `link u w Loop` as we did when we checked for inclusion.

Let us test the new version of `LTL_Saturation_Strategy` with node equality based blocking. Consider the formula $\mathsf{F}P \wedge \neg P \wedge \mathsf{X}\neg P$. LoTREC returns the premodel that is shown in Fig. 7.7.

Let us further test this tableaux procedure for **LTL**. Consider the formulas $\mathsf{G}\mathsf{X}P$ and $\mathsf{G}(\mathsf{F}P \wedge \neg P)$. The former is satisfiable while the latter is unsatisfiable. Figure 7.8 shows the result of running LoTREC on $\mathsf{G}\mathsf{X}P$. Figure 7.9 shows the result of running it on $\mathsf{G}(\mathsf{F}P \wedge \neg P)$. Both execution traces terminate due to the node equality check. The open premodel of the former is extensible to an **LTL** model, whereas the only open premodel of the latter is an inextensible premodel, and the input formula is therefore **LTL** unsatisfiable.

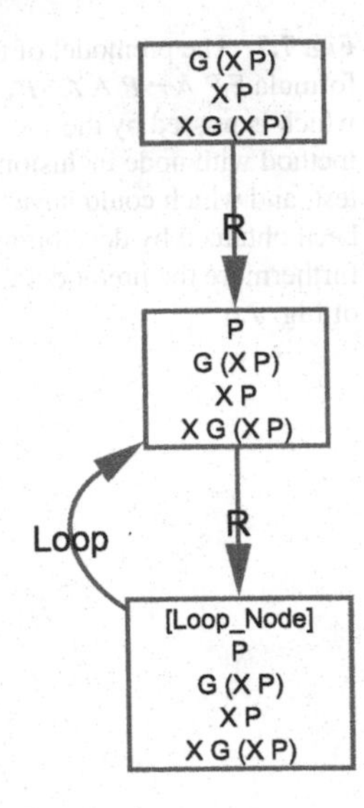

Fig. 7.8 Model construction for the formula $\mathsf{G}\mathsf{X}P$

7.1.6 LoTREC Strategy for LTL

The tableaux procedure for **LTL** works as follows.

1. We iterate the application of the rules for the boolean connectives and for the temporal operators, blocking nodes when their labels equal those of some ancestor node.
2. We check whether all eventualities are fulfilled.
3. We mark `Inextensible_Premodel` all those roots of premodels where there exist some unfulfilled eventualities.

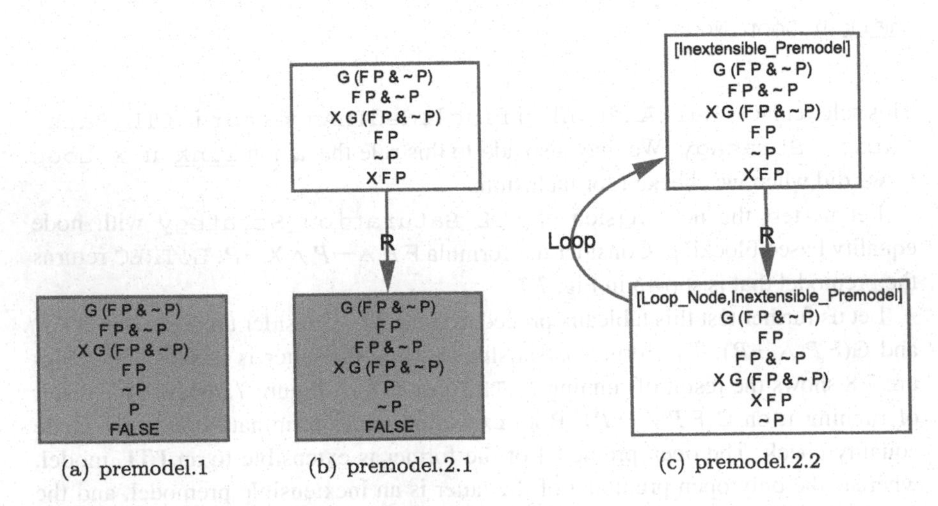

Fig. 7.9 Model construction for the formula $\mathsf{G}(\mathsf{F}P \wedge \neg P)$

The strategy calls the above `LTL_Saturation_Strategy` (where the rule `MarkAndLink_ifIncludedInAncestor` has been replaced by `Mark_ifEqualToAncestor`). Here it is:

```
Strategy LTL_Strategy
  LTL_Saturation_Strategy
  repeat
    Finally_fulfilledAtSameNode
    Finally_fulfilledAtDescendant
    Finally_fulfilledAtLoopAncestor
    Until_fulfilledAtSameWorld
    Until_fulfilledAtDescendant
    Until_fulfilledAtLoopAncestor
  end
  repeat
    Finally_unfulfilledIsInextensible
    Until_unfulfilledIsInextensible
    PropagateInextensibilityUp
    PropagateInextensibilityDown
  end
end
```

7.2 PDL: Propositional Dynamic Logic

Dynamic logics have not only a language of formulas but also a proper language of programs. Complex programs are built from atomic programs by means of the program connectives of sequential composition, nondeterministic composition, test, and iteration. (These are the standard program connectives of dynamic logic; extensions of that basic language with more program operators exist, such as converse and intersection.) The link between programs and formulas is provided by modal constructions of the form

$$[\pi]A$$

stating that after all the terminating executions of the program π, the property A holds. For instance, the formula [while $\neg(x=0)$ do $x := x-1]x=0$ expresses that after every terminating execution of the program while $\neg(x=0)$ do $x := x-1$, it holds that $x=0$.

Dynamic logics were introduced in the 1970s in theoretical computer science by Vaughn Pratt, Dexter Kozen, Rohit Parikh, and others as a modal logic to reason about programs [HKT00]. They subsume their predecessors Hoare logic [HKT00] and algorithmic logic [MS87]. The latter provides constructions of the form

$$Pre\{\pi\}Post$$

Table 7.1 The connectives of **PDL**

Operator	Name	Reading
$X;Y$	sequential composition	do X then do Y
$X \cup Y$	nondeterministic choice	do either X or Y
$A?$	test	if A is true then continue, else fail
X^*	iteration	do X an arbitrary number of times

stating that if the precondition *Pre* is true, then after all executions of the program π, the postcondition *Post* is true. An example is

$$y = 1 \left\{\textsf{while} \neg(x=0) \textsf{ do } x := x - 1\right\} x = 0 \wedge y = 1.$$

This is translated into dynamic logic as follows:

$$Pre \rightarrow [\pi]Post.$$

For instance, $y = 1 \rightarrow [\textsf{while} \neg(x=0) \textsf{ do } x := x - 1](x = 0 \wedge y = 1)$. Dynamic logic was also used by philosophers such as Krister Segerberg in order to reason about events and actions [Seg92].

In propositional dynamic logic **PDL**, complex programs are not built from concrete atomic programs such as the above $x := x - 1$, but rather from *abstract atomic programs* from some set $\mathcal{I}_0$.

The tableaux method for **PDL** is a bit more complicated than that for **LTL**: for **LTL** we only checked the fulfilment of formulas of the form $\mathsf{F}A$ and $A\mathsf{U}B$, while for **PDL** we have to embed a full model checker.

7.2.1 Syntax of PDL

In dynamic logics the formula $[X]A$ is read "After every possible execution of program X, A is true." That program X might be complex. In our definition of formulas of Sect. 2.1.5, all the edge labels of modal operators are taken from a 'flat' set $\mathcal{I}$ that has no structure. The language of **PDL** therefore requires a new definition.

Let $\mathcal{P}$ be a set of atomic formulas and let $\mathcal{I}_0$ be a set of atomic programs. The elements of $\mathcal{I}_0$ are noted I, J, etc. Complex programs are then built from atomic programs by using connectives that are familiar from programming languages. We denote complex programs by X, Y, etc. The intended meaning of such complex programs is given in Table 7.1.

Observe that tests have formulas as arguments. Formally we therefore have to define the set of complex programs $\mathcal{I}$ and the set of complex formulas $\mathcal{F}or$ by mutual induction, as the smallest set such that:

- $\mathcal{I}_0 \subseteq \mathcal{I}$;
- if $X, Y \in \mathcal{I}$, then $X; Y \in \mathcal{I}$;
- if $X, Y \in \mathcal{I}$, then $X \cup Y \in \mathcal{I}$;
- if $A \in \mathcal{F}or$, then $A? \in \mathcal{I}$;
- if $X \in \mathcal{I}$, then $X^* \in \mathcal{I}$;
- $\mathcal{P} \subseteq \mathcal{F}or$;
- if $A \in \mathcal{F}or$, then $\neg A \in \mathcal{F}or$;
- if $A, B \in \mathcal{F}or$, then $A \wedge B \in \mathcal{F}or$;
- if $A, B \in \mathcal{F}or$, then $A \vee B \in \mathcal{F}or$;
- if $A, B \in \mathcal{F}or$, then $A \rightarrow B \in \mathcal{F}or$;
- if $A \in \mathcal{F}or$ and $X \in \mathcal{I}$, then $[X]A \in \mathcal{F}or$.

The language of programs captures the usual constructions in procedural programming languages. For example, the complex program

$$(A?; X) \cup (\neg A?; Y)$$

allows us to express "if A then X else Y," and the complex program

$$(A?; X)^*; \neg A?$$

allows us to express "while A do X." The formula $[X]\bot$ expresses that the program X is not executable.

7.2.2 Semantics of PDL

A **PDL** model is of the form $M = (W, R, V)$, where the set of possible worlds W, the accessibility relations $R(X) \subseteq W \times W$ (one per $X \in \mathcal{I}$), and the valuation $V : \mathcal{P} \longrightarrow 2^W$ are just as in multimodal logics, cf. Chap. 2. The truth conditions for the connectives are as in Definition 7.

However, not every model of the above kind is going to be a legal Kripke model of **PDL**. The reason is that up to now we have no guarantee that complex programs are interpreted as they should be. A *standard* **PDL** *model* has to satisfy the following constraints:

$$\begin{aligned}
R(X; Y) &= R(X) \circ R(Y) \\
&= \{\langle w, u\rangle \mid \text{there is } v \text{ such that } \langle w, v\rangle \in R(X) \text{ and } \langle v, u\rangle \in R(X)\} \\
R(X \cup Y) &= R(X) \cup R(Y) \\
R(A?) &= \{\langle w, w\rangle \mid M, w \Vdash A\} \\
R(X^*) &= (R(X))^* \\
&= \{\langle w, u\rangle \mid \text{there is } n, v_1, \ldots, v_n \text{ such that } v_1 = w,\ v_n = u, \text{ and } \\
&\qquad \langle v_i, v_{i+1}\rangle \in R(X) \text{ for all } i < n\}
\end{aligned}$$

The relation $R(X^*)$ is the reflexive and transitive closure of the relation $R(X)$.

In standard models it may happen that atomic programs are nondeterministic. One can exclude this by stipulating that for atomic programs I, the accessibility relation $R(I)$ must be deterministic: if $\langle w, u\rangle \in R(I)$ and $\langle w, v\rangle \in R(I)$ then $u = v$. The logic of such models is called Deterministic **PDL**, noted **DPDL**.

The following equivalences are valid in standard **PDL** models.

$$[X;Y]A \leftrightarrow [X][Y]A$$

$$\langle X;Y\rangle A \leftrightarrow \langle X\rangle\langle Y\rangle A$$

$$[X \cup Y]A \leftrightarrow [X]A \wedge [Y]A$$

$$\langle X \cup Y\rangle A \leftrightarrow \langle X\rangle A \vee \langle Y\rangle A$$

$$[A?]B \leftrightarrow \neg A \vee B$$

$$\langle A?\rangle B \leftrightarrow A \wedge B$$

$$[X^*]A \leftrightarrow A \wedge [X][X^*]A$$

$$\langle X^*\rangle A \leftrightarrow A \vee \langle X\rangle\langle X^*\rangle A$$

7.2.3 LoTREC Rules for PDL: Saturating Premodels

Just as for **LTL**, the first part of the tableaux procedure consists in saturating premodels by decomposing formulas. We have the following kinds of rules:

- Rules for the boolean connectives: they are those for modal logic **K** of Sect. 3.7, collected in the strategy `BooleanRules`;
- Rules for negations followed by modal operators, i.e., for formulas of the form $\neg[X]A$ and $\langle X\rangle A$: just as the rewriting rules of **LTL**, they are

```
Rule NotPos_rewrite
  hasElement w not pos variable X variable A

  add w nec variable X not variable A

Rule NotNec_rewrite
  hasElement w not nec variable X variable A

  add w pos variable X not variable A
```

- Rules for formulas of the form $\langle X\rangle A$ and $[X]A$, where X is a complex, non-atomic program: they rewrite these formulas according to the equivalences of Sect. 7.2.2;

- Rules for formulas of the form $\langle I \rangle A$ and $[I]A$ formulas, where $I \in \mathcal{I}_0$ is an *atomic program*: just as the standard rules for $\Box$ and $\Diamond$ of `K_Strategy`, they create new I-successors.

In the sequel we give the rules of the last two kinds.

Rules for Sequential Composition To deal with modal operators of the form $\langle X; Y \rangle$ and $[X; Y]$ we define the following two rules:

```
Rule Pos_Seq
  hasElement w pos seq variable X variable Y variable A

  add w pos variable X pos variable Y variable A
```

```
Rule Nec_Seq
  hasElement w nec seq variable X variable Y variable A

  add w nec variable X nec variable Y variable A
```

The first rule reduces every $\langle X; Y \rangle A$ formula to the formula $\langle X \rangle \langle Y \rangle A$, whereas the second reduces every $[X; Y]A$ to $[X][Y]A$.

Rules for Nondeterministic Composition According to the equivalences $\langle X \cup Y \rangle A \leftrightarrow \langle X \rangle A \vee \langle Y \rangle A$ and $[X \cup Y]A \leftrightarrow [X]A \wedge [Y]A$, we define the following two rules:

```
Rule Pos_Union
  hasElement w pos union variable X variable Y variable A

  add w or pos variable X variable A pos variable Y
    variable A
```

```
Rule Nec_Union
  hasElement w nec union variable X variable Y variable A

  add w and nec variable X variable A nec variable Y
    variable A
```

Rules for Test Rules for the test implement the equivalences $\langle A? \rangle B \leftrightarrow A \wedge B$ and $[A?]B \leftrightarrow \neg A \vee B$.

```
Rule Pos_Test
  hasElement w pos test variable A variable B

  add w and variable A variable B
```

```
Rule Nec_Test
  hasElement w nec test variable A variable B

  add w or not variable A variable B
```

Rules for Iteration A formula $\langle X^*\rangle A$ is true in a world w if and only if, either A is true at w or $\langle X\rangle\langle X^*\rangle A$ is. In the second case, we are *postponing* the fulfilment of $\langle X^*\rangle A$ to an X-successor of w. This uses the valid equivalence $\langle X^*\rangle A \leftrightarrow (A \vee \langle X\rangle\langle X^*\rangle A)$. For the dual $[X^*]A$ we similarly use the equivalence $[X^*]A \leftrightarrow (A \wedge [X][X^*]A)$. We therefore define the following two rules in LoTREC:

```
Rule Pos_Star
  hasElement w pos star variable X variable A

  duplicate premodel_postponing
  add w variable A
  add premodel_postponing.w pos star variable X variable A
```

```
Rule Nec_Star
  hasElement w nec star variable X variable A

  add w variable A
  add w nec variable X nec star variable X variable A
```

Rules for Modal Operators with Atomic Programs After the iteration of the above rules, every complex program of the form $\langle X\rangle A$ or $[X]A$ is reduced to either a formula of the form $\langle I\rangle B$ or $[I]B$, where I is an atomic program and B is some formula. We then call the following two rules.

```
Rule Pos_AtomicProgram
 hasElement w pos variable I variable A
 isAtomic variable I

 createNewNode u
 link w u variable I
 add u variable A
```

```
Rule Nec_AtomicProgram
 hasElement w nec variable I variable A
 isLinked w u variable I

 add u variable A
```

Observe that these two rules are exactly the same as the Pos and Nec rules for **K**.

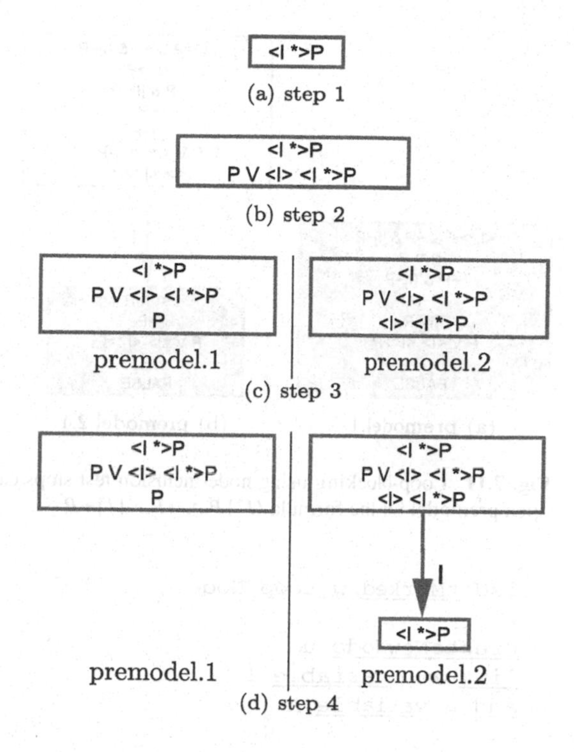

Fig. 7.10 Non terminating premodel construction for $\langle I^* \rangle P$

7.2.4 LoTREC Rules for PDL: Ensuring Termination

The above method does not terminate when dealing with a $\langle X^* \rangle$-formula. For instance, let us consider the model construction of the formula $\langle I^* \rangle P$, as shown in Fig. 7.10.

As the reader may notice, at step 4 the same initial node with the input formula occurs twice in premodel.2, which means that these steps will be repeated over and over again. So tableaux for $\Diamond X^* P$ behave in the same way as those for $\mathsf{F}A$ in **LTL**.

Just as in **LTL**, we may check for loops and block nodes, and just as we have seen in Sect. 7.1.5, it has to be a node equality test instead of a node inclusion test: model construction for the satisfiable formula $\langle I^* \rangle P \wedge \neg P \wedge [I]\neg P$ stops too early and the open premodel it returns is not extensible to a model, as illustrated in Fig. 7.11.

So we have to guarantee termination by blocking equal nodes. These nodes are tagged first by the rule `Mark_ifEqualToAncestor` that we have defined in Sect. 7.1.5 for **LTL**. We are then able to avoid further development of these nodes by adding to the node-creating rules the (negative) condition `isNotMarked w Loop_Node`. This gives us the following rule:

```
Rule Pos_AtomicProgram_ifUnblocked
  hasElement w pos variable I variable A
  isAtomic variable I
```

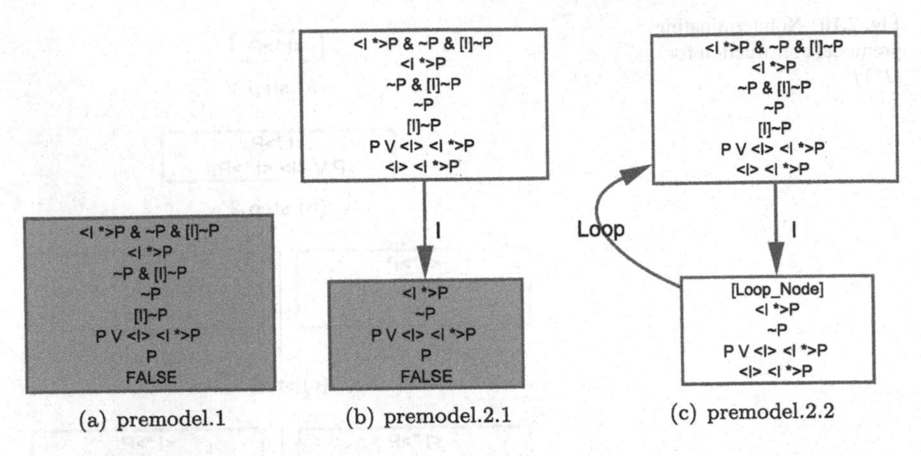

(a) premodel.1 (b) premodel.2.1 (c) premodel.2.2

Fig. 7.11 Loop-blocking using node-inclusion test stops earlier, and does not give an extensible open premodel for the formula $\langle I^*\rangle P \wedge \neg P \wedge [I]\neg P$

```
  isNotMarked w Loop_Node

  createNewNode u
  link w u variable I
  add u variable A
```

As for the strategy, we should call as long as possible all the rules for the boolean connectives and the rules for test, choice, sequence and iteration. After that we call the rule `Mark_ifEqualToAncestor`, and then we call `Pos_Atomic`. Such a strategy might be the following:

```
Strategy PDL_Saturation_Strategy
  repeat
    repeat
      repeat
        BooleanRules
      end
      NotNec_rewrite
      NotPos_rewrite
      Pos_Test
      Nec_Test
      Pos_Union
      Nec_Union
      Pos_Seq
      Nec_Seq
      Pos_Star
      Nec_Star
    end
    Mark_ifEqualToAncestor
    Pos_AtomicProgram_ifUnblocked
```

Fig. 7.12 An extensible open premodel for the formula $\langle I^* \rangle P \wedge \neg P \wedge [I]\neg P$

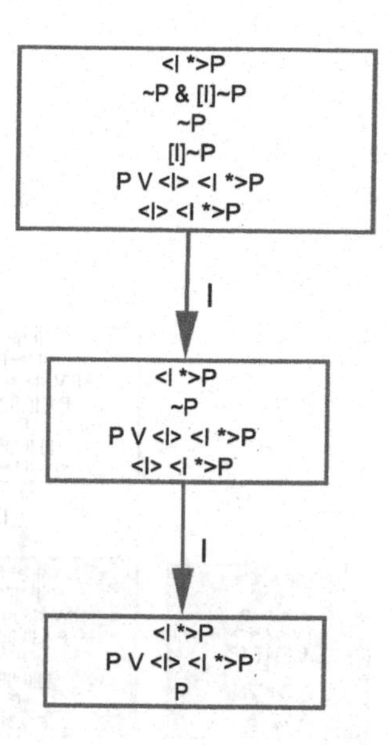


```
    Nec_AtomicProgram
  end
end
```

Running with this terminating tableaux procedure allows us to find an open premodel for the formula $\langle I^* \rangle P \wedge \neg P \wedge [I]\neg P$, as shown in Fig. 7.12.

7.2.5 LoTREC Rules for PDL: Fulfilment of Eventualities

Blocking loop nodes may lead to open premodels with unfulfilled `<X*>`-formulas. For instance, running `PDL_Saturation_Strategy` on the unsatisfiable formula $[I^*]P \wedge \langle I^* \rangle P$ yields three closed premodels and one open premodel, and the latter is reported as an *inextensible* premodel (non-extensible to a model) since `<I*> P` is not fulfilled in it (see Fig. 7.13).

We may think that we can decide whether an eventuality `<X*> A` is fulfilled at a node `w` by simply checking for the presence of the formula A in one of its successors, as it is the case in **LTL** (or other temporal logics). But this is not sufficient. In **PDL**, we must verify in addition that this successor is connected to `w` by a finite number of `R(X)` edges. However, `X` could be a complex program.

A first solution is to proceed *à la* Pratt [Pra80], by keeping track, for every eventuality `<X*> A` of the information about its *postponement* (potentially leading to its unfulfillment) along the `X`-paths built while treating the subformulas of this eventuality.

A second solution is to proceed *à la* de Giacomo & Massacci [dGM00] (who are inspired from model checking techniques used for μ-calculus). They rename each

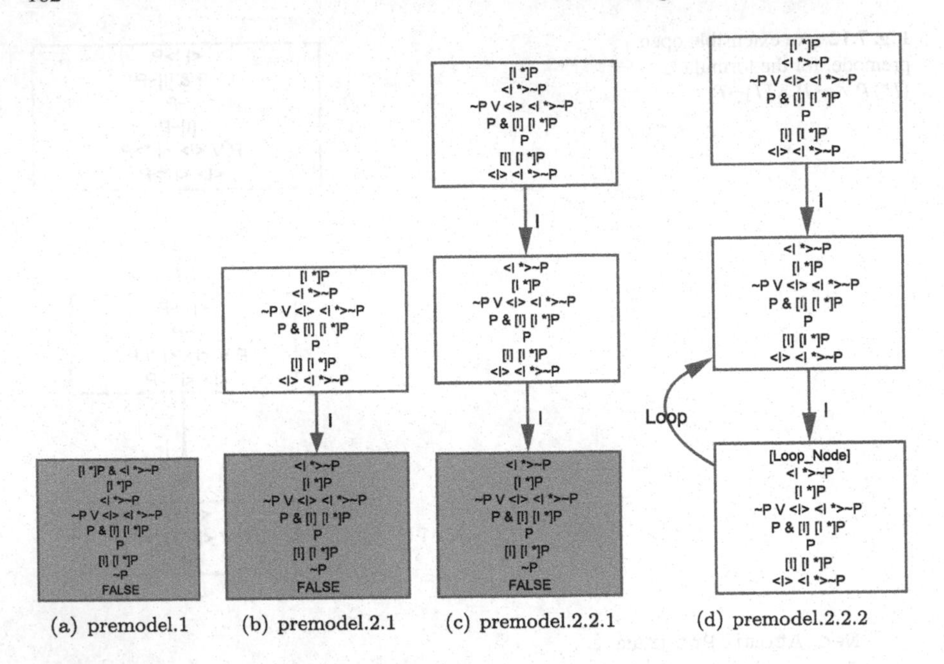

(a) premodel.1 (b) premodel.2.1 (c) premodel.2.2.1 (d) premodel.2.2.2

Fig. 7.13 An inextensible open premodel for the formula $[I^*]P \wedge \langle I^* \rangle P$

eventuality `<X*> A` in a given node `w` by an auxiliary variable, let us say `E`, and then reduce the `<X*> A` formula to `<X> E`. When `E` is found later in a successor node `u` of `w` then we know that `w` is connected to `u` through `X`-steps.

Here we introduce an alternative, simpler solution that can be implemented in LoTREC: to check every formula, including the eventualities, in a one-pass bottom-up verification. This solution keeps the premodels clearer by avoiding additional cumbersome edges or information in the graph structure of the premodels. It is however less efficient than the other two solutions that we have mentioned above.

Our procedure uses a tagging technique, just as the procedure that we have introduced in Chap. 6. The present model checking procedure is however not exactly the same since the premodels are not fully specified and since the analysis of the input formula was done on a different basis. In addition, in our model checking procedure we only tag formulas by `True` (and never by `False`).

Checking Literals Remember that a *literal* is either an atomic formula $P \in \mathcal{P}$ or the negation $\neg P$ of some $P \in \mathcal{P}$.

When there is a node where both literals `P` and `~P` coexist then `False` will be added to it by the `Stop` rule at some point. Hence the premodel is reported as closed and will not be developed further. It will therefore not be checked by the rules that we are going to define here.

Otherwise, an atomic formula `P` or its negation `~P` is supposed to be true when it is added by the rules of our procedure. Our rules therefore tag every positive literal `P` and negative literal `~P` occurring in a node by `True`.

```
Rule Mark_PositiveLiteral
 hasElement w variable P
 isAtomic variable P

 markExpressions w variable P True
```

```
Rule Mark_NegativeLiteral
 hasElement w not variable P
 isAtomic variable P

 markExpressions w not variable P True
```

Checking Classical Formulas As for classical rules, we define a rule to tag double negations formulas `~~A`, a rule to tag conjunctions and two rules for disjunctions, as follows:

```
Rule Mark_Not_Not
  hasElement w not not variable A
  isMarkedExpression w variable A True

  markExpressions w not not variable A True
```

```
Rule Mark_And
  hasElement w and variable A variable B
  isMarkedExpression w variable A True
  isMarkedExpression w variable B True

  markExpressions w and variable A variable B True
```

```
Rule Mark_Or_left
  hasElement w or variable A variable B
  isMarkedExpression w variable A True

  markExpressions w or variable A variable B True
```

```
Rule Mark_Or_right
  hasElement w or variable A variable B
  isMarkedExpression w variable B True

  markExpressions w or variable A variable B True
```

Checking Reduced Formulas In Sect. 7.2.3 we have rewritten formulas of the form `<X> A` and `[X] A` with complex programs `X`. For example, the rule `Pos_Seq` reduces every formula of the form `<X ; Y> A` to the formula `<X> <Y> A`. According to this decomposition, we tag `<X ; Y> A` by `True` in the nodes where the formula `<X> <Y> A` is tagged by `True`. Whence the rule:

```
Rule Mark_Pos_Seq
  hasElement w pos seq variable X variable Y variable A
  isMarkedExpression w
        pos variable X pos variable Y variable A True

  markExpressions w
        pos seq variable X variable Y variable A True
```

The rules tagging other kinds of reduced formulas, including classical formulas that are reduced to other formulas, are all defined in the same way as `Mark_Pos_Seq`. Let us call them `Mark_Pos_Seq`, `Mark_Nec_Seq`, `Mark_Pos_Union`, etc.

Remark 13 Reduced formulas are checked w.r.t. the way they are rewritten. For example, suppose that we want to define the rule `Mark_Not_Imp` which checks if a formula `~(A -> B)` should be tagged `True` at a given node `w`. Then this rule could be defined in two different ways, depending on how `~(A -> B)` was reduced by the rule `Not_Imp` during the model construction:

- if the formulas `A` and `~B` are added to `w` then `Mark_Not_Imp` should tag it `True` whenever both `A` and `~B` are tagged so;
- otherwise, if it was rewritten to the formula `A & ~B` then the rule `Mark_Not_Imp` should tag the formula `~(A -> B)` by `True` whenever the formula `A & ~B` is tagged so.

The first version of the rules is:

```
Rule Mark_Not_Imp
 hasElement w not imp variable A variable B
 isMarkedExpression w variable A True
 isMarkedExpression w not variable B True

 markExpressions w not imp variable A variable B True
```

whereas the second version is:

```
Rule Mark_Not_Imp
 hasElement w not imp variable A variable B
 isMarkedExpression w and variable A not variable B True

 markExpressions w not imp variable A variable B True
```

So the LoTREC user should define the checking rules according to how his reduction rules are defined.

Checking <I> A and [I] A Formulas When the program X is atomic then formulas of the form <I> A are easily checked by the following rule:

```
Rule Mark_Pos_AtomicProgram
 hasElement w pos variable I variable A
 isMarkedExpression u variable A True
 isLinked w u variable I

 markExpressions w pos variable I variable A True
```

Formulas of the form [X] A are checked exactly as the []-formulas in the standard model checking procedure of Chap. 6, viz. by checking if A is tagged True in all children nodes.

```
Rule Mark_Nec_AtomicProgram
 hasElement w nec variable I variable A
 isMarkedExpressionInAllChildren w variable A variable I True

 markExpressions w nec variable I variable A True
```

Inheriting Tags from Loop-Parent Nodes At a loop node w whose labels equal those of some ancestor, the formulas of the form <X> A cannot be tagged True by the above rules because w is blocked and has no further successors. We therefore copy the True tag of every formula in an ancestor node to the loop-node:

```
Rule Mark_FormulasInLoopNodes
 hasElement w variable A
 isLinked w u Loop
 isMarkedExpression u variable A True

 markExpressions w variable A True
```

With this rule, we come to the end of our model checking rules. These rules should be called repeatedly at the end of the **PDL** strategy that constructs the pre-models. Here is a strategy for doing that.

Here is the strategy for putting all the rules together.

```
Strategy PDL_Fulfilment_Checking_Strategy
  repeat
    Mark_PositiveLiteral
    Mark_NegativeLiteral
    Mark_NotNot
    Mark_And
    Mark_Or_left
    Mark_Or_right
    Mark_Pos_Seq
```

```
    Mark_Nec_Seq
    Mark_Pos_Union
    Mark_Nec_Union
    Mark_Pos_Test
    Mark_Nec_Test
    Mark_Pos_Iteration
    Mark_Nec_Iteration
    Mark_Pos_AtomicProgram
    Mark_Nec_AtomicProgram
    Mark_FormulasInLoopNodes
  end
end
```

7.2.6 LoTREC Strategy for PDL

We now have almost all the rules we need for **PDL**. Here are two further rules which improve readability of LoTREC's output premodels.

Once the application of the above rules is finished, we are sure that only unfulfilled formulas are not tagged True in our open premodels. Hence, we can report inextensible premodels by performing the following simple check:

```
Rule Eventuality_unfulfilledIsInextensible
 hasElement w pos star variable X variable A
 isNotMarkedExpression w pos star variable X variable A True

 mark w Inextensible_Premodel
```

If we furthermore want to display this information in every node of an inextensible premodel then we may use the rules defined at the end of Sect. 7.2.5 and propagate the Inextensible_Premodel tag to all the nodes of an inextensible premodels.

Here is the strategy putting all the rules together.

```
Strategy PDL_Strategy
  PDL_Saturation_Strategy
  PDL_Fulfilment_Checking_Strategy
  Eventuality_unfulfilledIsInextensible
end
```

If we run PDL_Strategy on the formula $[I^*]P \wedge \langle I^* \rangle P$ then the open premodel.2.2.2 of Fig. 7.13 is reported as an inextensible premodel. This is shown in Fig. 7.14.

It is not the case that every open premodel with a loop is an inextensible premodel. For example, running PDL_Strategy on $[I^*](\langle I^* \rangle P \wedge \langle I \rangle \neg P)$ returns some closed premodels, some inextensible open premodels, and an open premodel that is stopped due to the loop test and that is extensible to a model of the input formula. The latter premodel is shown in Fig. 7.15.

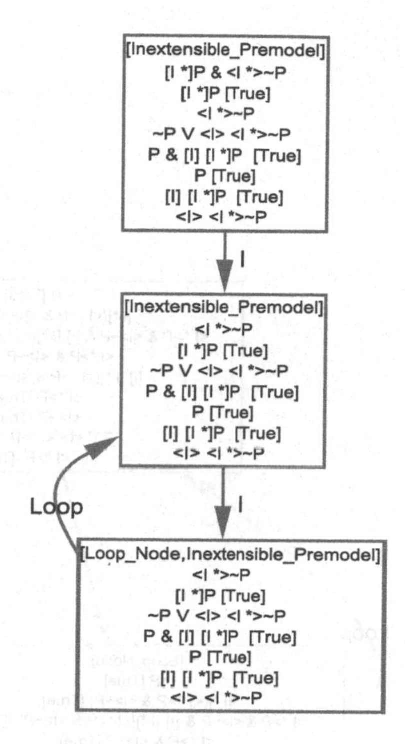

Fig. 7.14 Inextensible open premodels are designated due to model checking

If we remove the complex formulas from this premodel and only keep the atomic formulas then we find the nice little model that is depicted in Fig. 7.16.

Exercise 81 Prove that the instance $(P \wedge [I^*](P \rightarrow [I]P)) \rightarrow [I^*]P$ of the so-called induction axiom is valid in **PDL**.

Exercise 82 Implement a tableaux procedure for **DPDL**, i.e., Deterministic **PDL**. Prove that the equivalences $[X](P \vee Q) \leftrightarrow [X]P \vee [X]Q$ and $\langle X \rangle (P \wedge Q) \leftrightarrow \langle X \rangle P \wedge \langle X \rangle Q$ are valid in **DPDL**.

Exercise 83 As we have mentioned in the introduction of the present chapter, in the logic of *common knowledge* the set $\mathcal{I}_0$ is viewed as a set of agents and there is not only an accessibility relation $R(I)$ for each individual agent I in the set of agents $\mathcal{I}_0$, but also accessibility relations $R(S)$ for each non-empty set of agents $S \subseteq \mathcal{I}_0$. When S is different from a non-singleton then $R(S)$ is defined to be the reflexive and transitive closure of the union of the set of individual relations. Formally:

$$R(S) = \left(\bigcup_{I \in S} R(I) \right)^*$$

Implement a tableaux proof procedure for the logic of common knowledge.

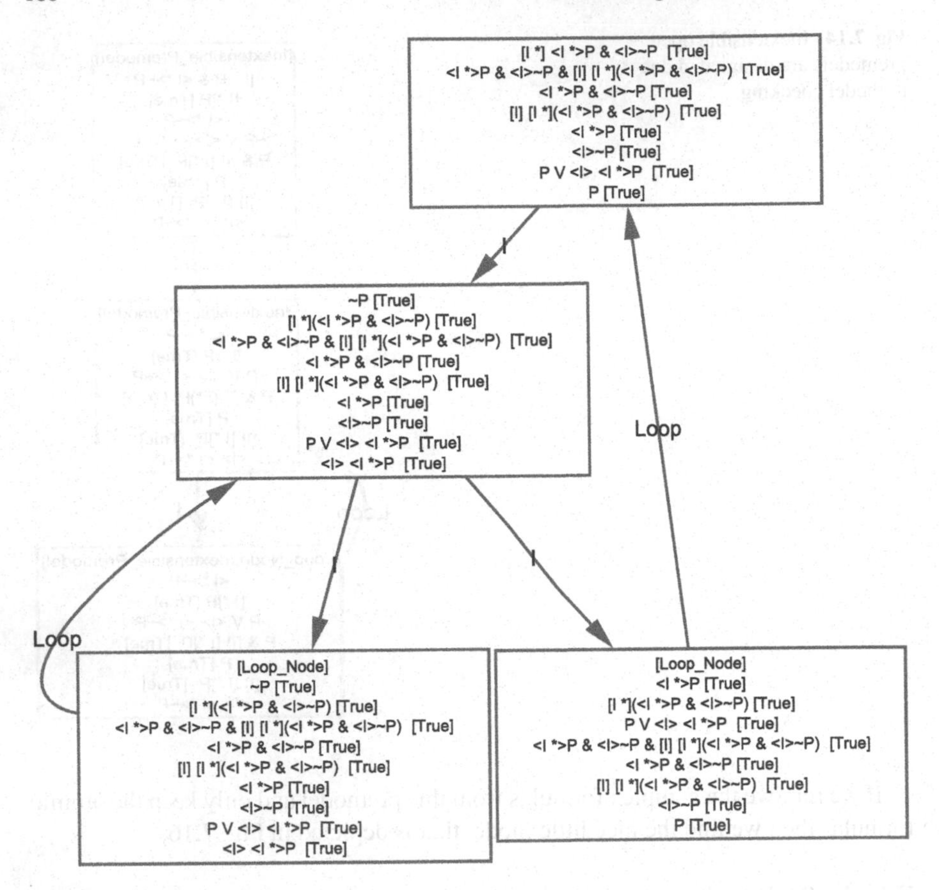

Fig. 7.15 Open premodel involving loops could be a good premodel

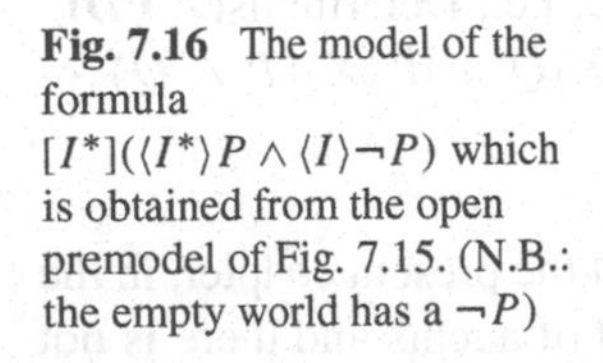

Fig. 7.16 The model of the formula $[I^*](\langle I^*\rangle P \wedge \langle I\rangle\neg P)$ which is obtained from the open premodel of Fig. 7.15. (N.B.: the empty world has a $\neg P$)

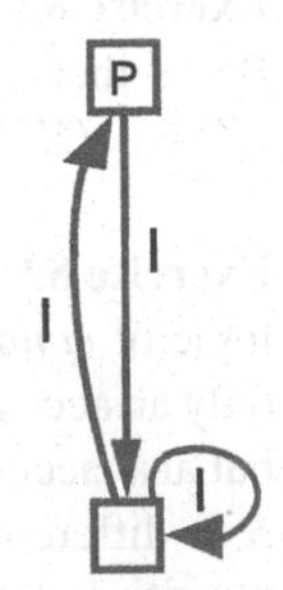

Exercise 84 The logic of *common belief* has accessibility relations $R(S)$ for each non-empty set of agents $S \subseteq \mathcal{I}_0$. When S is different from a non-singleton then the accessibility relation $R(S)$ is the transitive closure of the union of the set of

individual relations. This is written formally as follows:

$$R(S) = \left(\bigcup_{I \in S} R(I) \right)^{+}$$

So the difference with common knowledge is that $R(S)$ is not necessarily reflexive.
Implement a tableaux proof procedure for the logic of common belief.

7.3 Conclusion

In this chapter we handled two logics with transitive closure: **LTL** and **PDL**. Their semantics are quite different, but share the same kind of formulas expressing 'eventually': $\mathsf{F}A$ in **LTL** and $\langle I^* \rangle A$ in **PDL**. Hence, their model construction methods share the same rule of 'eventuality postponement:' if A is eventually true then A is true at this step, otherwise at the next step we have to check whether Awill be eventually true.

In addition, both procedures need a node-equality loop check, otherwise the method may stop before giving the chance to some eventualities to be satisfied. When the procedure halts then a model checking procedure has to be performed in order to check whether every eventuality is fulfilled.

Our implementation of **LTL** and **PDL** in LoTREC cannot compete with existing optimised implementations for these logics.

References

[AG09] P. Abate and R. Goré. The tableau workbench. *Electronic Notes in Theoretical Computer Science*, 231:55–67, 2009.

[AtC06] C. Areces and B. ten Cate. Hybrid logics. In P. Blackburn, J. van Benthem, and F. Wolter, editors, *Handbook of Modal Logic, volume 3*. Elsevier Science, Amsterdam, 2006.

[AvBN98] H. Andréka, J.F.A.K. van Benthem, and I. Nemeti. Modal languages and bounded fragments of predicate logic. *Journal of Philosophical Logic*, 27(3):217–274, 1998.

[BCM+03] F. Baader, D. Calvanese, D.L. McGuinness, D. Nardi, and P.F. Patel-Schneider, editors. *Description Logic Handbook*. Cambridge University Press, Cambridge, 2003.

[BBW06] P. Blackburn, J.F.A.K. van Benthem, and F. Wolter. *Handbook of Modal Logic*, volume 3 of *Studies in Logic and Practical Reasoning*. Elsevier Science, New York, 2006.

[BdRV01] P. Blackburn, M. de Rijke, and Y. Venema. *Modal Logic. Cambridge Tracts in Theoretical Computer Science*. Cambridge University Press, Cambridge, 2001.

[Bet55] E.W. Beth. Semantic entailment and formal derivability. In *Mededlingen van de Koninklijke Nederlandse Akademie van Wetenschappen, Afdeling Letterkunde, N.R, volume 18(3)*, pages 309–342, 1955. Reprinted in Jaakko Intikka (ed.) The Philosophy of Mathematics, Oxford University Press, 1969.

[BK08] C. Baier and J.-P. Katoen. *Principles of Model Checking. Representation and Mind Series*. The MIT Press, Cambridge, 2008.

[BS00] F. Baader and U. Sattler. Tableau algorithms for description logics. *Studia Logica*, 69:2001, 2000.

[CFdCGH97] M.A. Castilho, L. Fariñas del Cerro, O. Gasquet, and A. Herzig. Modal tableaux with propagation rules and structural rules. *Fundamenta Informaticae*, 32(3/4):281–297, 1997.

[Che80] B.F. Chellas. *Modal Logic. An Introduction*. Cambridge University Press, Cambridge, 1980.

[CL90a] P.R. Cohen and H.J. Levesque. Intention is choice with commitment. *Artificial Intelligence*, 42(2–3):213-261, 1990.

[CL90b] P.R. Cohen and H.J. Levesque. Persistence, intentions, and commitment. In P.R. Cohen, J. Morgan, and M.E. Pollack, editors. *Intentions in Communication*, Chapter 3, pages 33–69. MIT Press, Cambridge, 1990.

[Coo71] S.A. Cook. The complexity of theorem proving procedures. In *Proceedings Third Annual ACM Symposium on Theory of Computing*, pages 151–158. ACM, New York, 1971.

[CP09] W.A. Carnielli and C. Pizzi. *Modalities and Multimodalities. Logic, Epistemology, and the Unity of Science*. Springer, Berlin, 2009.

O. Gasquet et al., *Kripke's Worlds*, Studies in Universal Logic,
DOI 10.1007/978-3-7643-8504-0, © Springer Basel AG 2014

[CZ97] A. Chagrov and M. Zakharyaschev. *Modal Logic. Oxford Logic Guides*. Oxford University Press, London, 1997.

[dGM00] G. de Giacomo and F. Massacci. Combining deduction and model checking into tableaux and algorithms for converse-PDL. *Information and Computation*, 162(1/2):117–137, 2000.

[FdCGHS05] L. Fariñas del Cerro, O. Gasquet, A. Herzig, and M. Sahade. Modal tableaux: completeness vs. termination. In S. Artemov, H. Barringer, A. d'Avila Garcez, L.C. Lamb, and J. Woods, editors. *We Will Show Them! Essays in Honour of Dov Gabbay, volume 1*, pages 587–614. College Publications, London, 2005.

[FGV05] M. Fisher, D.M. Gabbay, and L. Vila. *Handbook of Temporal Reasoning in Artificial Intelligence*. Elsevier Science, New York, 2005.

[FHMV95] R. Fagin, J.Y. Halpern, Y. Moses, and M.Y. Vardi. *Reasoning About Knowledge*. MIT Press, Cambridge, 1995.

[Fit83] M. Fitting. *Proof Methods for Modal and Intuitionistic Logics*. D. Reidel, Dordrecht, 1983.

[Gab02] D.M. Gabbay. *Labelled Deductive Systems, volume 1*. OUP, London, 2002.

[GHS06] O. Gasquet, A. Herzig, and M. Sahade. Terminating modal tableaux with simple completeness proof. In G. Governatori, I. Hodkinson, and Y. Venema, editors, *Advances in Modal Logic, volume 6*, pages 167–186. King's College Publications, London, 2006.

[GKWZ03] D.M. Gabbay, A. Kurucz, F. Wolter, and M. Zakharyaschev. *Many-Dimensional Modal Logics: Theory and Applications*, volume 148 of *Studies in Logic and the Foundations of Mathematics*. Elsevier, Amsterdam, 2003.

[Gor99] R. Goré. Tableau methods for modal and temporal logics. In M. D'Agostino, D.M. Gabbay, R. Hähnle, and J. Posegga, editors, *Handbook of Tableau Methods*, pages 297–396. Hingham, MA, USA. Kluwer Academic, Dordrecht, 1999.

[GSS03] D.M. Gabbay, D. Skvortsov, and V. Shehtman. *Quantification in Nonclassical Logic. Volume 1*, volume 153 of *Studies in Logic and the Foundations of Mathematics*. Elsevier, Amsterdam, 2003.

[HC68] G.E. Hughes and M.J. Cresswell. *An Introduction to Modal Logic*. Methuen, London, 1968.

[HC84] G.E. Hughes and M.J. Cresswell. *A Companion to Modal Logic*. Methuen, London, 1984.

[Hin62] J.K.K. Hintikka. *Knowledge and Belief*. Cornell University Press, Ithaca, 1962.

[HKT00] D. Harel, D. Kozen, and J. Tiuryn. *Dynamic Logic*. The MIT Press, Cambridge, 2000.

[Hor98a] I. Horrocks. The fact system. In H. de Swart, editor, *Proc. of the 2nd Int. Conf. on Analytic Tableaux and Related Methods (TABLEAUX'98)*, volume 1397 of *Lecture Notes in Artificial Intelligence*, pages 307–312. Springer, Berlin, 1998.

[Hor98b] I. Horrocks. Using an expressive description logic: FaCT or fiction. In *Proc. of the 6th Int. Conf. on Principles of Knowledge Representation and Reasoning (KR'98)*, pages 636–647, 1998.

[HSZ96] A. Heuerding, M. Seyfried, and H. Zimmermann. Efficient loop-check for backward proof search in some non-classical propositional logics. In *TABLEAUX '96: Proceedings of the 5th International Workshop on Theorem Proving with Analytic Tableaux and Related Methods*, London, UK, pages 210–225, Springer, Berlin, 1996.

[Kle67] S.C. Kleene. *Mathematical Logic*. Wiley, New York, 1967.

[Kri63] S. Kripke. Semantical analysis of modal logic. *Zeitschrift für Mathematische Logik und Grundlagen der Mathematik*, 9:67–96, 1963.

[Len78] W. Lenzen. *Recent Work in Epistemic Logic*. North Holland, Amsterdam, 1978.

[Len95] W. Lenzen. On the semantics and pragmatics of epistemic attitudes. In A. Laux and H. Wansing, editors, *Knowledge and Belief in Philosophy and AI*, pages 181–197. Akademie Verlag, Berlin, 1995.

[Lew73] D. Lewis. *Counterfactuals*. Basil Blackwell, Oxford, 1973.

[LL32] C.I. Lewis and C.H. Langford. *Symbolic Logic*. Century Company, London, 1932. Reprinted, New York: Dover Publications, 2nd edition, 1959.

[Mas00] F. Massacci. Single step tableaux for modal logics. *Journal of Automated Reasoning*, 24(3):319–364, 2000.

[MS87] G. Mirkowska and A. Salwicki. *Algorithmic Logic*. D. Reidel, Dordrecht, 1987.

[NPW02] T. Nipkow, L.C. Paulson, and M. Wenzel. *Isabelle/HOL—A Proof Assistant for Higher-Order Logic*, volume 2283 of *LNCS*. Springer, Berlin, 2002.

[Pau89] L.C. Paulson. The foundation of a generic theorem prover, *Journal of Automated Reasoning*, 5, 1989.

[Pea00] J. Pearl. *Causality: Models, Reasoning, and Inference*. Cambridge University Press, New York, 2000.

[Pra76] V.R. Pratt. Semantical considerations on Floyd-Hoare logic. In *Proceedings of the 17th IEEE Symposium on Foundations of Computer Science*, pages 109–121, 1976.

[Pra80] V.R. Pratt. A near-optimal method for reasoning about action. *Journal of Computer and System Sciences*, 20(2):231–254, 1980.

[Pri57] A.N. Prior. *Time and Modality*. Oxford University Press, Oxford, 1957.

[Pri67] A.N. Prior. *Past, Present, and Future*. Oxford University Press, Oxford, 1967.

[Seg92] K. Segerberg. Getting started: beginnings in the logic of action. *Studia Logica*, 51:347–378, 1992.

[Smu95] R. Smullyan. *First Order-Logic*. Dover, New York, 1995. First print: 1968.

[Sol76] R.M. Solovay. Provability interpretations of modal logic. *Israel Journal of Mathematics*, 25:287–304, 1976.

[Sta68] R. Stalnaker. A theory of conditionals. In *Studies in Logical Theory*, volume 2 of *American Philosophical Quarterly (Monograph Series)*, pages 98–112. Blackwell, Oxford, 1968 (reprinted in E. Sosa, ed., Causation and Conditionals. Oxford University Press, 1975; reprinted in Harper, W.L. and Stalnaker, R. and Pearce, G., eds., Ifs. Reidel, Dordrecht, 1981; reprinted in W. L. Harper and B. Skyrms, eds., Causation in decision, belief change and statistics, Vol.2. Reidel, Dordrecht, 1988, pp 105–134; reprinted in F. Jackson, ed., Conditionals. Oxford University Press, Oxford Readings in Philosophy, 1991).

[TH06] D. Tsarkov and I. Horrocks. Fact++ description logic reasoner: system description. In *Proc. of the Int. Joint Conf. on Automated Reasoning (IJCAR 2006)*, volume 4130 of *Lecture Notes in Artificial Intelligence*, pages 292–297. Springer, Berlin, 2006.

[TSK12] D. Tishkovsky, R.A. Schmidt, and M. Khodadadi. The tableau prover generator mettel2. In *Proc. of the Eur. Conf. on Logic in Artificial Intelligence (JELIA 2012)*, volume 7519 of *Lecture Notes in Artificial Intelligence*. Springer, Berlin, 2012.

[Var96] M.Y. Vardi. Why is modal logic so robustly decidable. In N. Immerman and P.G. Kolaitis, editors, *Descriptive Complexity and Finite Models*, volume 31 of *DIMACS Series in Discrete Mathematics and Theoretical Computer Science*, pages 149–184. American Mathematical Society, Providence, 1996.

[Ver10] R. (L.C.) Verbrugge. Provability logic. In E.N. Zalta, editor, *The Stanford Encyclopedia of Philosophy*. Winter 2010 edition, 2010.

[vW51] G.H. von Wright. Deontic logic. *Mind*, 60:1–15, 1951.

[Woo02] M. Wooldridge. *An Introduction to Multiagent Systems*. Wiley, New York, 2002.

Index

O. Gasquet et al., *Kripke's Worlds*, Studies in Universal Logic,
DOI 10.1007/978-3-7643-8504-0, © Springer Basel AG 2014